T0226900

STILL GOING WRONG!

Case Histories of Process Plant Disasters and
How They Could Have Been Avoided

STILL GOING WRONG!

Case Histories of Process Plant Disasters and How They Could Have Been Avoided

Trevor Kletz

ELSEVIER

AMSTERDAM • BOSTON • HEIDELBERG • LONDON
NEW YORK • OXFORD • PARIS • SAN DIEGO
SAN FRANCISCO • SINGAPORE • SYDNEY • TOKYO

Gulf Professional Publishing is an imprint of Elsevier

G | P
P | 🛡

Gulf Professional Publishing is an imprint of Elsevier.

Copyright © 2003 by Butterworth–Heinemann

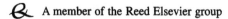 A member of the Reed Elsevier group

Library of Congress Cataloging-in-Publication Data
Kletz, Trevor A.
    Still going wrong! : case histories of process plant disasters and how they could
have been avoided / by Trevor A. Kletz.
      p. cm.
    Includes index.
    ISBN-13: 978-0-7506-7709-7   ISBN 0-7506-7709-0 (acid-free paper)
    1. Chemical plants — Accidents. I. Title.

    TP155.5.K539 2003
    660'.2804 — dc21
    ISBN-13: 978-0-7506-7709-7   2003051440
    ISBN-10: 0-7506-7709-0

British Library Cataloguing-in-Publication Data
A catalogue record for this book is available from the British Library.

The publisher offers special discounts on bulk orders of this book.
For information, please contact:

Manager of Special Sales
Butterworth–Heinemann
200 Wheeler Road
Burlington, MA 01803
Tel: 781-904-2500
Fax: 781-221-1615

For information on all Butterworth–Heinemann publications available, contact our
World Wide Web home page at: http://www.bh.com

Transferred to Digital Printing 2009

Contents

Forethoughts

Some years ago I went to a conference at which a newly appointed director of safety began his presentation with the assertion that "safety management is not rocket science." And he was right. Rocket science is a trivial pursuit compared to the management of safety. There are only a limited number of fuel types capable of lifting a payload into space; but the variety of ways in which harm can come to people is legion. Although writing a procedure to achieve some productive aim is not easy, particularly when the task is complex, it is always possible. In contrast, there are not enough trees in the rainforest to support all the procedures necessary to guarantee a person's safety while performing that activity. . . . James Reason

Several years ago after reading *What Went Wrong*, I realized I could use it to "wake up" my people to the dangers and horror others have experienced. All of the line supervisors and managers were given copies of the book, and every month, during our regular meetings, each was to talk about something from the book that could happen here, and what we needed to do to be sure it didn't. Not only was this educational and motivational, but it also was a way to get people to discuss and share feelings of vulnerability (something not easily articulated by many of this breed). . . . Shelley Roth, operations manager of a chemical plant

Human language is a spectacular mechanism for transferring ideas from one mind to another, allowing us to accumulate knowledge over many generations. . . . Daniel Hillis

Introduction

During my time as an industrial safety adviser, I published a monthly Safety Newsletter, which described incidents that had occurred in the company and elsewhere and the actions needed to prevent them from happening again. By the time I retired, I was sending out 2,000 copies every month, many outside the company, to regulators, academics, and other companies, and in return I received reports on other accidents. After retiring from industry, I published a selection of extracts from the Safety Newsletter in a book called *What Went Wrong? Case Histories of Process Plant Disasters*. It is now in its 4th edition, is twice as long as the first edition, and contains reports on many more recent incidents. It has sold more copies than any of my other books. I had the good fortune to work for a company, Imperial Chemical Industries (ICI), that was at the time exceptionally willing to share its safety information with other organizations. It also gave me considerable freedom to take the actions I thought necessary to improve the safety record, which had deteriorated in the decade before my appointment. My book, *By Accident — A Life Preventing Them in Industry* (PV Publications, London, 2001), relates my experiences and shows how I developed the views described in this book.

Unfortunately, there is no shortage of new accident reports and this book contains a selection of them, from many sources, including little-known publications and private communications from companies and individuals. The incidents described are not "carbon copies" of those in *What Went Wrong?* (referred to in the text hereafter as *WWW*) as I have chosen those that illustrate new causes as well as familiar ones. This book therefore complements *WWW* and does not replace it. However, there is one difference. In *WWW*, I emphasized the immediate technical causes and the changes in design and methods of working needed to prevent them from happening again. In this book I have, whenever possible, also discussed the underlying weaknesses in the management systems. It is not possible to do this in every case as the information is not always available. Too many reports still describe only the immediate technical causes. I do not blame their authors for this. Most of them are too close to the "coal-face." They want to solve the immediate technical problems and get the plant back on line in a safe manner as soon as they can, so they concentrate on the immediate technical causes. Before approving the reports, more senior people should look for the underlying weaknesses that result in poor designs, poor work methods, failures to learn from the past, tendencies to blame people who make occasional but inevitable errors, and so on. They should also see that changes that cannot be made on the existing plants are fed back to the design organizations, both in-house and contractors, for future use.

In *WWW*, some of the chapters covered different types of equipment while others covered procedures such as maintenance or modifications. In this book most of the chapters cover procedures but a number of reports on explosions and leaks are collected under these headings. This book also emphasizes the multiple causes of accidents. As a result the accidents described in the chapter on the management of change, for example, also have other causes while some incidents in other chapters also involve the management of change. Similarly, several scattered reports show that some accidents cannot be prevented by more detailed instructions but only by giving people a better understanding of the process. This makes the allocation of incidents to chapters rather arbitrary so I have included many cross references and a comprehensive index. Missing from this book is a chapter on human error. This is because all accidents are due to human error. Someone, usually a manager or supervisor, has to decide what to do; someone, often a designer, has to decide how to do it; someone, usually an operator, has to do it. All may make errors, of different sorts, for different reasons. Human errors are too diverse to be treated as a single group, and I find it useful to classify them as follows:

MISTAKES: They occur because someone does not know what to do. To prevent them we need better training or instructions or changes to the plant design or work method so that the task is easier.

VIOLATIONS OR NONCOMPLIANCE: They occur when someone knows what to do but decides not to do it. We need to explain why the correct methods should be followed and we need to check that they are followed and not turn a blind eye. We should also see if the task can be made easier.

MISMATCHES: The job is beyond the mental or physical ability of the person asked to do it, perhaps beyond anyone's ability. We should change the design or work method.

SLIPS AND LAPSES OF ATTENTION: These are inevitable from time to time and so we should change designs or work methods so as to remove opportunities for human error.

This classification is discussed in more detail in Chapter 16. An underlying principle behind this book is that whenever possible we should remove situations that are error prone rather than expect people never to make errors. There is more about human error in my book, *An Engineer's View of Human Error*, 3rd edition (Taylor & Francis, Philadelphia, PA and Institution of Chemical Engineers, Rugby, UK, 2001). Another feature illustrated in almost every chapter is the way that the lessons of past accidents are soon forgotten (or never learned) and the accidents are allowed to happen again (see Section 16.10).

As in my Safety Newsletter and in *WWW*, I have preserved the anonymity of the companies where the accidents occurred, except when this is mentioned in the title of a published report. When no reference is cited, the information came from a private communication or my own experience.

This book is intended for all who work in industry, especially the chemical, oil, and other process industries, and are involved in production, maintenance, or design, at any level or in any capacity. I hope I am not being immodest when I say that the most senior people with responsibilities for production, operators, and everyone in between can learn something from it.

If I describe an incident that occurred at your plant, you may notice that one or two details have been changed. I may have done this to make a complicated story simpler but without affecting the essential message or — and this is more likely — the incident did not occur at your plant at all. Rather, another plant had a similar incident.

My advice is given in good faith but without warranty. You should satisfy yourself that it applies to your particular circumstances. You may feel that some of my recommendations go too far, or not far enough, or that other solutions are more appropriate. Fair enough, but please do not ignore the accidents. They have happened and could happen again. Do not say, "We have systems to prevent them" unless you are sure that they are always followed, everywhere, all the time. Perhaps your systems are followed most of the time but someone turns a blind eye when a job is urgent. Also remember that all systems have limitations. All systems can do is make the most of people's knowledge and experience by applying these in a systematic way. If people lack knowledge and experience, the system is an empty shell.

If you decide on a course of action, try not to yield to pressure, obstacles, complacency, or example ("we always do it this way."), but do yield to sound technical arguments. I used to tell my safety colleagues in industry that a job was not finished when they gave their advice. It was not finished until their advice was followed or they were convinced by technical arguments that it should be changed. In science it is permissible to say that we do not know the answer to a problem, but this is not possible in plant design and operation. We have to make a decision even though the evidence is conflicting. To quote David Pye, "It is quite impossible for any design to be the logical outcome of the requirements simply because the requirements being in conflict, their logical outcome is an impossibility." Information on what has gone wrong in the past can help us find the best balance between these conflicting requirements.

Many of the incidents I describe did not have serious results. By good fortune no one was killed or injured and damage was slight. For example, a leak of flammable liquid did not ignite or corrosion was spotted in time. Do not ignore these incidents. Next time you may not be so lucky. In the following pages, I criticize the performance of some organizations. However, I am not suggesting that they neglected safety to save money. A few may have done so but the vast majority did not. Most accidents occur because the people in charge do not see the hazards (what could occur) or underestimate the risk (the probability that it will occur), because they do not know what more could be done to remove the hazards or reduce the risk, or because they allow standards of performance to slip, all common human failings. Unexpected results are far more common than conspiracies to cut costs.

Since the first edition of *WWW* was published in 1985, the press and public have become more likely to look for someone to blame when something goes wrong (see Section 14.9). The old legal principle of "no liability without fault" is being replaced by "those in charge should pay compensation whether or not they were negligent." This increases the pressure for better safety but it also makes some companies reluctant to publish all the facts, even internally, so that others can learn from them. With this newer principle, there may be no net gain.

HOW TO USE THIS BOOK

1. Read it straight through. Ask if each incident could occur in your plant and, if so, write down what you intend to do to prevent it from happening.

2. Dip into it at odd moments or use it to pick a topic for discussion at meetings, an item to include in your safety bulletin, or something to look for during plant audits or inspections. The chapters can be read in any order.

3. Refer to it when you become interested in something new as a result of changes in responsibility or new problems. However, this book covers only limited aspects of process safety. For a comprehensive treatment of the subject, see F. P. Lees' *Loss Prevention in the Process Industries*, 2nd edition, (Butterworth-Heinemann, Woburn, MA and Oxford, UK, 1996).

4. Use the book to train new employees, at all levels, so that they realize what can happen when people do not follow recognized procedures or good practice.

5. If you are a teacher, use the book to show your students why accidents occur and to illustrate the relevance of their studies to life in industry.

6. Put copies of the book, open at the appropriate page, on the desks of people who have allowed any of the incidents described to happen. Perhaps they will read the book and avoid unnecessary incidents in the future.

A high price was paid for the information in this book: people were killed or injured and billions of dollars worth of equipment was damaged. Someone has paid the "tuition fees." There is no need for you to pay them again.

ACKNOWLEDGMENT

I would like to thank the many friends and colleagues, past and present, who provided information for this book, both reports on incidents and comments on their causes and ways of preventing them from happening again.

By the time this book, my 11th, is published, I shall be in my 82nd year. It is my final harvest.

Trevor Kletz

A NOTE ON NOMENCLATURE

Different words are used, in different countries, to describe the same job or piece of equipment. Some of the principal differences between the United States and the United Kingdom are listed here. Within each country, however, there are differences between companies.

Management Terms

Job	US name	UK name
Operator of plant	Operator	Process worker
Maintenance worker	Craftsman or mechanic	Fitter, electrician, etc.
Operator in charge of others	Lead operator	Chargehand, Assistant foreman or junior supervisor
Highest level normally reached by promotion from operator	Foreman	Foreman or supervisor
First level of professional management (usually in charge of a single unit)	Supervisor	Plant manager
Second level of professional management	Superintendent	Section or area manager
Senior manager in charge of site containing many units	Plant manager	Works manager

The different meanings of the terms "supervisor" and "plant manager" in the US and UK should be noted. In this book I have usually used the US names but have avoided the term "plant manager" as it has such different meanings in the two countries and have described the professional person in charge of a single unit, or group of small units, as the unit manager. The terms "foreman" and "lead operator" are understood in both countries, though the use of the former is now becoming outdated in the UK. I have used "supervisor" where the term could refer to any type of supervisor. Some US-owned companies in the UK use US names, others use UK names or a mixture of the two. See the table.

Chemical Engineering Terms

Certain items of plant equipment have different names in the two countries. Some common examples are:

US	UK
Accumulator	Reflux drum
Agitator	Mixer or stirrer
Air masks	Breathing apparatus (BA)
Blind or spade	Slip-plate
Carrier	Refrigeration plant
Cascading effects	Knock-on or domino effects
Check valve	Non-return valve
Clogged (of filter)	Blinded
Consensus standard	Code of practice
Conservation vent	Pressure/vacuum valve
Dike, berm	Bund
Discharge valve	Delivery valve
Division (in electrical area classification)	Zone
Downspout	Downcomer
Expansion joint	Bellows
Explosion proof	Flameproof
Faucet	Tap
Fiberglass reinforced plastic (FRP)	Glass reinforced plastic (GRP)
Figure 8 plate	Spectacle plate
Flame arrester	Flame trap
Flashlight	Torch
Float (of rotameter)	Bobbin
Fractionation	Distillation
Gauging (of tanks)	Dipping
Gasoline	Petrol
Generator	Dynamo or alternator
Ground	Earth
Horizontal cylindrical tank	Bullet
Hydro (Canada)	Electricity
Install	Fit
Insulation	Lagging
Interlock*	Trip**

*In the U.K., "interlock" is used to describe a device that prevents someone opening from one valve while another is open (or closed).

**"Trip" describes an automatic device that closes (or opens) a valve when a temperature, pressure, flow, etc. reaches a preset value.

US	UK
Inventory	Stock
Lift-truck	Fork lift truck
Loading rack	Gantry
Manway	Manhole
Mill water	Cooling water
Nozzle	Branch
OSHA (Occupational Health & Safety Administration)	HSE (Health & Safety Executive)
Pedestal, pier	Plinth
Pipe diameter (internal)	Bore
Pipe rack	Pipebridge
Plugged	Choked
Rent	Hire
Rupture disc or frangible	Bursting disc
Scrutinize	Vet
Seized (of a valve)	Stuck shut
Shutdown	Permanently shut down
Sieve tray	Perforated plate
Siphon tube	Dip tube
Spade	Slip-plate
Sparger or sparge pump	Spray nozzle
Spigot	Tap
Spool piece	Bobbin piece
Stack	Chimney
Stator	Armature
Tank car	Rail tanker
Tank truck	Road tanker
Torch	Cutting or welding torch
Tower	Column
Tow motor	Fork lift truck
Tray	Plate
Turnaround	Shutdown
Utility hole	Manhole
Valve cheater	Wheel dog
Water seal	Lute
Wrench	Spanner
C-wrench	Adjustable spanner
Written note	Chit
$M	Thousand dollars
$MM	Million dollars
STP	60 °F, 1 atmosphere
32 °C, 1 atmosphere	STP
NTP	32 °C, 1 atmosphere

I have usually used the US terms, but habits are hard to break and some UK terms have crept in. Measurements are usually in SI units followed by imperial units in brackets. However, pipe diameters are usually in inches only as it seems overly pedantic to describe the familiar 1-in pipe as a 25-mm one. Gallons are US gallons (1 UK gallon = 1.2 US gallons).

Chapter 1

Maintenance

People should have to take a class on this information before they receive their undergraduate degrees in engineering. Nobody really tells us this stuff. . . . — A message from a chemical engineering student who found *What Went Wrong?* in a library

The longest chapter in my other book of case histories, *What Went Wrong?* (*WWW*), is Chapter 1, Preparation for Maintenance. This is still the source of many accidents and more are described in the following pages. They have been chosen to emphasize that the need for good practice and its enforcement is as great as ever and to draw attention to features not discussed in *WWW*.

1.1 INADEQUATE PREPARATION ON A DISTANT PLANT

This accident occurred in a large, responsible international company but in a distant plant many thousands of miles away from the US and Europe. Pipework connected to a tank that had contained a flammable liquid was being modified. The tank was "washed clean with water," to quote the report. The foreman checked that the tank looked clean and that there was no smell. The valves on the tank and the manway cover were all closed, or so it was thought, and a permit was issued for welding on the pipework. One of the pipes was cut with a hacksaw and a section removed. When a welder started to weld the replacement section an explosion occurred in the tank. The welder was hit by the manway cover and hurled 5 m (16 ft) to the ground. He died from his injuries.

1.1.1 What Went Wrong?

- Water washing may remove all the liquid from a tank but it cannot remove all the vapor. Tests for flammable vapor should have been carried out INSIDE AND OUTSIDE the tank before work started, and it is good practice to place a portable gas detector alarm near the welding site in case conditions change.

- Two of the valves on the tank were found to be open to atmosphere. One of them, on the top of the tank, probably provided the flame to ignition path as the welder was working several meters away. The foreman should have checked these valves before issuing the permit-to-work.

- It is possible that the valve between the tank and the line being welded was leaking. The lines between the tank and the welding operations should have been blinded.

- The job was completed by removing all the pipework and modifying it in the workshop. That could have been done before the explosion.

- The procedures for preparing equipment for maintenance were grossly inadequate or ignored (or both). It is most unlikely that this was the first time that a job had been prepared in such a slipshod way, and more senior and professional staff should have noticed what was going on.

Before you say, "This couldn't happen in my company," remember Bhopal (or Longford; see Section 4.2). Do you know what goes on in your overseas plants or in that little faraway plant that you recently acquired, not because you really wanted it but as part of a larger deal? Do you circulate all your recommended practices and accident reports to these outstations? Do you audit their activities?

1.2 PRECAUTIONS RELAXED TOO SOON

When a whole unit is shut down for an extended overhaul, the usual practice is to isolate the unit at the battery limits by inserting blinds in all pipe lines, to remove all hazardous materials, and to check that any remaining concentrations are low enough for safety. Many publications [1] describe how this can be done. It is then not necessary to isolate individually every piece of equipment that is going to be inspected or maintained. (However, equipment that is going to be entered should still be individually isolated by blinding or disconnection.)

After a long shutdown, there is obviously a desire to get back on line as soon as possible. A few jobs are not quite finished. Can we remove the battery limit isolations, or some of them, and start warming up a section of the plant where all the work is complete?

The correct answer is "Yes, but first the equipment that is still being worked on must be individually blinded. Do not depend on valve isolations. Valves can leak" (see Section 12.3). The following incident occurred because this advice was not followed.

A fluid coker was starting up after a 4-wk shutdown. Work on some items of equipment, including the main fractionation column was not quite finished and its vent line was still open to the atmosphere. Some, but not all, of the lines leading to this column were blinded to support this work so it was decided to start removing the battery limit blinds. When the blind on the low pressure natural gas supply line was removed, passing gas was detected in the plant, as the natural gas isolation valve was leaking. The blind was replaced but removed the next day. The leak then seemed small. Six hours later there was an explosion in the fractionation column. The trays were displaced and damaged but the shell was unharmed.

The precise route by which the gas got into the column is uncertain and is not described in the report [2]. It probably came from the leaking valve just described. However, the next level of cause is clear: before the battery limit blinds were removed, every line leading to equipment that was still being worked on or was open to the atmosphere should have been individually blinded. The underlying causes were taking chances to get the plant back on line quickly, and insufficient appreciation of the hazards.

1.2.1 Lessons Learned

Under this heading the report describes with commendable frankness some well-known information that was apparently not known to those in charge.

A SMALL QUANTITY OF FLAMMABLE GAS OR VAPOR CAN CAUSE A LARGE EXPLOSION WITH SEVERE CONSEQUENCES especially when the fuel is confined. As little as 5–15 kg (10–30 lb) of methane could have caused the damage as it is not necessary for the whole of the vessel to be filled with the flammable mixture. Vessels should be inerted if there is any possibility of flammable gas entering, through leaking valves or in other ways (but it is better to prevent gas entering by adequate blinding). To bring home to people the power of hydrocarbons, remind them that a gallon (4 liters, ≈3 kg or 1.5 lb), burned in a rather inefficient engine, can accelerate a ton of car to 70 mph (110 km/h) and push it 30 mi (50 km). Looked at this way, the damage to the column seems less surprising. Most of us get practical experience of the energy in hydrocarbons every day, but we do not relate it to the hydrocarbons we handle at work.

The quantity of gas that might leak through a closed valve is significantly more than most people realize.

CONSIDERATION SHOULD BE GIVEN TO DOUBLE BLOCKS AND A DRAIN ON BLINDING INSTALLATIONS THAT ARE TROUBLESOME OR HIGHLY SENSITIVE TO LEAKAGE. Note that the double block and drain (or vent) does not remove the need for more positive isolation by blinding. It makes the fitting of blinds safer and is adequate for quick jobs, but not for extended ones such as turnarounds.

VALVES ARE THE MAINSTAY OF ANY PLANT. Trying to stop leaks with excessive torque will damage them. Any that are troublesome should be noted for change at the next shutdown.

THE RIGOR WITH WHICH COMMISSIONING ACTIVITIES ARE CARRIED OUT IS OFTEN LESS THAN THAT WHICH IS APPLIED TO NORMAL OPERATING PROCEDURES . . . ALL PROCEDURES SHOULD BE WRITTEN IN A CLEAR, CONCISE AND CONSISTENT MANNER.

1.3 FAILURE TO ISOLATE RESULTS IN A FIRE

In the last incident, the equipment under maintenance was not isolated from a source of danger, natural gas, because blinds were removed prematurely and the consequences not thought through. In this incident, there was not only a leaking valve but no blinds were (or could be) inserted.

A pin-hole leak occurred on a 6-in diameter naphtha draw-off line from a fractionation column at a height of 34 m (112 ft) above ground level. Many attempts were made to isolate and drain the line but without success as the valve between the line and the column was passing intermittently when it was supposed to be closed and the bottom of the line was plugged with debris. Nevertheless, it was decided to replace a corroded 30 m

(100 ft) length of it with the plant on line, despite the fact that the workers doing so would be working at a height, with limited means of escape, and with hot pipework nearby. This decision was made at operator level and professional staff were not involved.

Two cuts were made in the pipe with a pneumatic saw. When naphtha leaked from the second cut, it was decided to open a flange and drain the line. As the line was being drained, there was a sudden release of naphtha from the first cut. It was ignited, probably by the hot surface of the column, and quickly engulfed the column. Four men were killed and another seriously injured.

The immediate cause of the fire was the grossly unsafe method of working. The plant should have been shut down. (If the line had been narrower and not corroded, it might have been possible to run a new line alongside the existing one and carry out an under-pressure connection.)

The underlying causes were:

• The technical and managerial staff were rarely seen on the site, did not take sufficient interest in the details of plant operation and, in particular, allowed an operator to authorize and control an obviously hazardous job.

• Employees at all levels had a poor understanding of the hazards.

• They did not recognize the need for a systematic evaluation of the hazards of specific jobs and the need to prepare a detailed plan of work [3].

In cases like this, managers have been known to say afterwards, "I didn't know this sort of thing was going on. If I had known, I would have stopped it." This is a poor excuse. It is a manager's job to know what is going on and this knowledge cannot be learned by sitting in an office, but by visiting the site, carrying out audits, and generally keeping one's eyes open. When an accident discloses a poor state of affairs, it is stretching credulity too far to claim that it was the first time that risks or shortcuts had been taken. They are usually taken many times before the result is an accident.

1.4 UNINTENTIONAL ISOLATION

Many incidents have occurred because someone isolating a flow or an electricity supply has not realized that he or she was also isolating the supply to other equipment besides the equipment intended for isolation. If this is not obvious from the position of the isolation valve, then a label should indicate which equipment or unit is supplied via the valve. Similarly, labels on fuse boxes and main switches should indicate which equipment or unit is supplied.

The flow of compressed air to a sampling system was isolated unintentionally. This was not discovered for some time as the bulb in the alarm light had failed. The operator canceled the audible alarm but with no indicator light to remind him he forgot that the alarm had sounded, or perhaps he assumed that flow had been restored. The alarm was checked weekly to make sure that the set point was correct but the alarm light was not checked.

Sometimes an unintentional isolation is the result of a slip. An operator was asked to switch a spare transformer on line in place of the working one. This was done remotely from the computer in the control room. He inadvertently isolated the working transformer before switching on the spare one. He realized his error almost immediately and the supply was restored within a minute. The report on the incident blamed distraction:

> It is apparent that the Control room is used as a gathering area for personnel, as well as a general thoroughfare for persons moving about the building, to the detriment of the Control room operator's concentration.

The report also suggested greater formality in preparing and following instructions when equipment is changed over. Though not suggested in the report, it should be simple for the computer program, when the computer is asked to isolate a transformer, to display a warning message such as, "Are you sure you want to shut down the electricity supply?" We get such messages on our computers when we wish to delete a file. There is no need for control programs to be less user-friendly than word processors.

Notice that the default action of the people who wrote the report was to describe ways of changing the operator's behavior rather than to look for ways of changing designs or methods of working (see Chapter 5).

1.5 BAD PRACTICE AND POOR DETAILED DESIGN

A reciprocating air compressor was shut down for repair. The process foreman closed the suction and delivery valves and isolated the electricity supply. He then tried to vent any pressure left in the machine by the method normally used, opening the drain valve on the bottom of the pulsation damper. It was seized and he could not move it. So instead he vented the pressure by operating the unloading devices on both cylinders. Unfortunately, this did not vent all parts of the machine, though the foreman and most of the workforce did not realize this. He then issued a permit-to-work for the repair of the machine.

Two men started to dismantle the machine. They noticed that the handle of the drain valve on the pulsation damper was vertical and assumed that the valve was open. They therefore assumed that the pressure had been blown off. After they had unbolted one component, it flew off, injuring one of them.

We can learn a number of lessons from this incident:

- Members of process teams often do not always understand in detail the construction of mechanical equipment or the way it works. They should therefore be given detailed instructions on the action to be taken when preparing such equipment for maintenance and, of course, encouraged to learn more about the equipment they operate (see also Section 1.9).

- When handing over the permit to the maintenance worker or foreman, the process foreman should have explained exactly what he had done. The report [4] does not state whether or not he handed it over in person but if he had done so he would presumably have mentioned that he was unable to blow off the pressure in the usual way. Unfortunately, it is all too common for people to leave permits on a table for others to pick up. This is bad practice. When permits are being issued or handed back on completion of a job, this should be done person-to-person.

- It is possible on many cocks and ball valves to remove the handle and turn it through to 90 degrees before replacing it. Such valves are accidents waiting to happen. We should use valves that tell us at a glance whether they are open or shut. Rising spindle valves are better than those with nonrising spindles. Ball valves and cocks should have handles that cannot be replaced wrongly.

- Drain valves often become plugged with scale or dirt. Valves used to blow off pressure should therefore be on the top rather than the bottom of a vessel.

- People dismantling equipment should always assume that it may contain trapped pressure and should proceed cautiously. They should loosen all bolts and prize the joint open so that any trapped pressure can blow off through the crack or, if the leak is serious, the joint can be remade.

1.6 DISMANTLING

1.6.1 Wrong Joint Broken

A supervisor decided to remove a number of redundant pipes and branches from a service trench. They had not been used for many years and were rusty and unsightly. An experienced man went around with a spray can and marked with green paint the sections of pipe that were to be removed and then a permit-to-work was issued to remove the marked sections.

One of the sections was a short vertical length of pipe, 75 mm (3 in) diameter and ≈1.5 m (5 ft) long, sticking up above a compressed air main that was still in use (see Figure 1-1). The valve between the pipe and the main was tagged to show that it was closed in order to protect equipment under maintenance. The short length of pipe was marked with several green patches. Unfortunately there was also some green paint on the flange below the isolation valve. This green paint might have been the remains of an earlier job or it might have accidentally got onto the flange while the pipe above it was being sprayed. The mechanic who had been asked to remove the pipe broke this flange. There was a sudden release of compressed air at a gauge pressure of ≈7 bar (100 psig). Fortunately the mechanic escaped injury.

The mechanic did not, of course, realize that the compressed air line was still in use. Like the old pipes he was removing, it was rusty and he assumed it was out of use.

This incident displays several examples of poor practice. Each job should have its own permit-to-work, which should make it quite clear which joint or joints should be broken. The report on the incident stated that in the future, maintenance workers should be shown precisely which piece of equipment is to be maintained, which joint to break, etc. However,

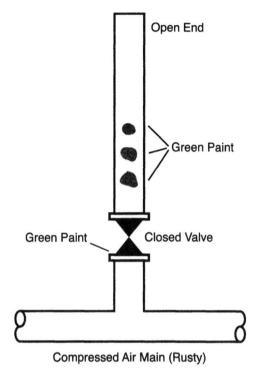

Open End

Green Paint

Green Paint   Closed Valve

Compressed Air Main (Rusty)

Figure 1-1. The lower flange was unbolted in error.

experience shows that this is not enough. Before starting work, the maintenance worker may go for his tools or for spares and then come back and break the wrong joint, or remove the wrong valve. Equipment to be worked on should be numbered or labeled and the number or name put on the permit-to-work. If there is no permanent number, a numbered tag should be tied to the flange that is to be broken, the valve that is to be changed, or the pipe that is to be cut (and at the point at which it has to be cut).

These identifying tags should be quite distinct in appearance from tags used to show that valves are isolating equipment under repair. For this purpose, padlocks and chains, or other locking devices, are better than tags as they prevent the valve being opened in error.

During the investigation, someone suggested that the job did not need a permit-to-work as it was noninvasive. However, the purposes of a permit-to-work are to define precisely the work to be done, list the hazards that are present, and describe the necessary precautions. If a job is not defined precisely, it may become invasive.

Though it did not contribute to the accident, rainwater will have collected above the valve and caused corrosion. The open end should have been blanked.

1.6.2 Trapped Pressure in Disused Equipment

Equipment that is no longer used or is not going to be used for some time should be emptied and made safe. If you are sure it will not be needed again, dismantle it as soon as you can. Leave knowledge, not problems, for your successors. They may not know what was in the line or how to handle it. Nevertheless, if you have to dismantle old equipment, do not assume that it has been made safe. Assume that pressure may be trapped behind solid plugs. Here are some examples:

(a) An old disused pipeline was being dismantled by cutting it into lengths with a hacksaw and lowering them to the ground. Both ends of the pipeline were open. When a mechanic cut into the pipe, a spray of sulfuric acid hit him in the face. Fortunately he was wearing goggles. There were two closed valves in the line, but no one had noticed them. The acid had attacked the metal, forming hydrogen, which pressurized the line [5].

(b) A stainless steel pipe was isolated at both ends and left for six years. After this time no one remembered what it was last used for. One end was still connected to the plant; the other, lower, end had a valve fitted on it. A mechanic was asked to dismantle the pipe. He opened this valve. Nothing came out. He then unbolted the joint between the valve and the pipe and prized the flanges apart. A little liquid dribbled out. He prized the flanges further apart. A large and forceful escape of gas, liquid and dirt sprayed the fitter and his assistant. The pipe had contained acetic acid and over the years it had corroded the pipe sufficiently to produce a pressure of hydrogen.

(c) A unit was "abandoned in place" to save the cost of demolition. A pump that handled a 50% solution of caustic soda was isolated by closing both valves but was not drained and the fuses were not removed. A contractor was asked to switch on a ventilation fan that served an adjoining area. The switch was next to that for the caustic pump, though 15 m away from the pump and the labels on the switches were very small. The contractor switched on the caustic pump in error. It ran between closed valves and overheated. There was a loud boom, which rattled windows 60 m (200 ft) away. The pump was damaged and dislodged from its baseplate [6].

There were five elementary errors. The incident would not have occurred if one of the following five tasks had been carried out:

- The pump had been drained.
- The pump had been defused.
- The switches were near the equipment they served. (Additional switches for emergency use could have been provided some distance away.)
- The labels were easy to read.
- Someone familiar with the unit had been asked to switch on the ventilation fan.

Incidentally, the same source describes how two other pumps were damaged because they were operated while isolated. One was switched on remotely; the casing was split into two. A power failure caused the other to stop. An operator closed the isolation valves, not realizing that the pump would restart automatically when power was restored. When it did, a bit of the pump was found 120 m (400 ft) away and local damage was extensive.

1.7 COMMISSIONING

A new unit was being commissioned. It was bigger and more complex than any of the other units on the site so the project and engineering teams had checked and double-checked everything, or so they thought. To make sure there were no leaks and that the instruments worked correctly they operated the plant with water, except for a vessel that was intended for the storage of a water-sensitive reagent. To avoid contaminating this vessel, it was left isolated by two closed valves, a manual valve on the vessel and a control valve below it.

This vessel was later filled with the reagent, and commissioning started. When an operator, standing on a ladder, opened the manual valve a cloud of dense white fumes surrounded him. Fortunately he was able to close the valve and escape without injury. There was no gasket in one of the joints between the two valves [7].

As it was impracticable to leak test this part of the unit with water, it should have been tested with compressed air, by either checking whether or not it retained the pressure or checking for leaks with soapy water. Note that during such a leak test with compressed air, the design pressure of equipment should not be exceeded. Pressure tests to check the integrity of equipment should normally be carried out with water — or other liquid — as then much less energy is released if the equipment fails.

Valves that are operated only occasionally, say, once per year at a planned turnaround, may be operable only from a ladder but valves that may be required during process upsets, such as leaks, should be easy to reach.

1.8 OTHER HIDDEN HAZARDS

In new plants, and extensions to old ones, we often find welding rods, tools, and odd bits of metal left in pipework. Even small bits of rubbish can harm machinery and most companies make sure that pipes are clean before new equipment is started up. The most extreme example of unwanted contamination occurred on a U.S. refinery when a length of new pipe, complete with plastic end caps, was being prepared for installation. Welders fitted a bend on one end of the pipe and then, with the end cap still in place, cut a length off the other end. They then found a propane cylinder just inches away from the cut (see Figure 1-2). Had it been a few inches nearer the end, there would have been a very nasty accident [8].

Therefore, one never knows what suppliers and construction teams have left inside new equipment. Have a good look before boxing it up or working on it for the first time (see also Section 1.10).

1.9 CHANGES IN PROCEDURE

An instrument probe in a tank truck used to carry gasoline had to be replaced. The job was done regularly by an experienced mechanic. After the new probe had been inserted, some electrical connections were made and secured with a heat-activated shrink-wrap sleeve. A propane torch was used to seal the shrink-wrap. This was hardly the most suitable tool to use on a vessel containing flammable vapor but as the probe was always

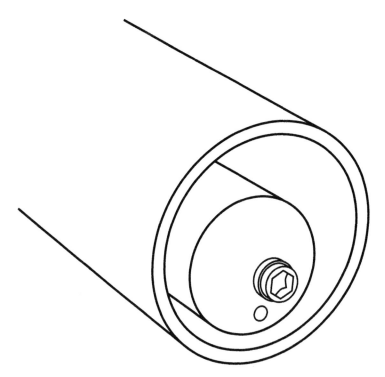

Figure 1-2. When a length of new pipe was cut open, a propane cylinder was found inside it. Fortunately the cutting tool just missed it.

replaced before the shrink-wrap was fitted, the vapor was not open to the atmosphere.

One day the tank truck was wanted back in use as soon as possible so the mechanic used a different type of electrical connection that could be made more quickly but required the application of heat. The mechanic used the torch; he had been using it for several years and the only difference, as he probably saw it, was that he was now using it at a slightly earlier stage of the job, before the probe had been replaced and while the tank was still open to the atmosphere. It exploded and killed the mechanic [9].

The mechanic and his supervisor did not understand the hazards and thought they were making only a minor change in the way the job was done. Many maintenance people do not understand process hazards (just as many process people do not understand equipment hazards; see Section

1.5). The company that owned or used the tank truck should have removed all flammable vapor from the tank before sending it for repair or at the very least should have made clear the nature of the hazards and the precautions to be taken.

When equipment is owned by one company, rented by another, and repaired by a third, responsibilities for maintenance, pressure testing, and inspection should be agreed upon and made clear.

1.10 DEAD-ENDS

Dead-ends in pipes have caused many pipe failures. Traces of water can accumulate in them and freeze or corrosive materials can dissolve in them (see *WWW*, Section 9.1.1). Other materials can also accumulate in them and remain there when the rest of the equipment has been emptied for maintenance (or for any other reason).

1.10.1 A Disused Pipe Becomes a Dead-end

A furnace was taken out of use. The 10-in diameter pipe that supplied coke oven gas to the burners was disconnected at the lower (furnace) end and closed by a valve, but the other end was left connected to the main. Ten years later a crack appeared in the top flange of the valve and gas leaked out. The crack was probably caused by the freezing of water that had collected in the pipe. Water and other liquids were normally removed from the main coke oven gas line via a number of drain lines but missing insulation had allowed the water in these lines to freeze. As an immediate measure, a blind was inserted immediately above the cracked valve. While this was being done, some tar oozed from the pipe.

A few days later a maintenance team started to replace the cracked valve with a blank flange fitted with a drain. When the flange was loosened the valve and blind dropped down several inches while hanging on the remaining bolts. A large amount of liquid sprayed out and soaked three of the workmen. It was ignited either by an infrared lamp used to warm the line or by a gas-fired space heater. Poor access and junk lying around prevented a quick escape by the workers; two were killed — one of them fell over the junk — and another was seriously injured [10].

There were at least eight things wrong and putting just one of them right could have prevented the accident.

- The redundant pipe should have been removed when it became clear that it was no longer needed. It is not good practice to leave disused pipes in position in the belief that one day someone might find a use for them.

- If this was not possible for any reason, the pipe should have been regularly drained to remove any liquid that accumulated. This should have been through a small valve fitted below or in place of the 10-in valve but not so small that it would be liable to choke, perhaps of 2 in internal diameter, but smaller for less viscous drainings. (The liquid should, of course, be drained into a closed container, not into a bucket and not onto the floor.)

- The missing insulation on the lines that drained the main coke oven gas lines should have been replaced.

- No sources of heat should have been allowed anywhere near equipment that might contain flammable material.

- All employees should have been told about other incidents in which liquids in coke oven gas lines had caught fire, including several that had occurred only days before as a result of the freezing of the drain lines. Many employees thought these liquids were not flammable.

- There should have been regular surveys of the unit to look for dead-end pipes, missing insulation, and other defects.

- The material that leaked out when the blind was being fitted should have been checked for flammability.

- The job should have been properly planned. The company's procedures were frequently ignored.

The dead and injured might have escaped had there been less junk lying around.

1.10.2 A Dead-End Inside a Vessel

Paint had to be removed from the manway of a reactor that had contained ethylene oxide. The reactor was swept out with nitrogen and tests showed that no oxygen or combustible gas or vapor could be detected. Unfortunately, the people who prepared the reactor overlooked a disused line on the base of the reactor that was permanently blinded. Some ethylene oxide that accumulated in this line evaporated and was ignited by sparks from the grinder used to remove the paint. A flash fire killed the man using the grinder but there was no explosion.

Why was the ethylene oxide not detected by the tests? According to the report, the sample tube used was 3.4 m (11 ft) long, not quite long enough to reach to the bottom of the reactor which was 3.7 m (12 ft) deep. In addition, the ethylene oxide may have been at the bottom of the disused line and as its vapor is heavier than nitrogen, it would diffuse out only slowly. There are other possibilities not mentioned in the report [7]. The sample tube might have absorbed the ethylene oxide so that it never reached the detector head (see Sections 2.4 and 10.4); and combustible gas detectors will not detect flammable gas unless air is present. The report does not say how much time elapsed between the tests and the grinding. A test at 8 AM, say, does not prove that the plant is still safe several hours later. Tests should be carried out just before work starts, and it is good practice to use a portable gas detector alarm, which gives an audible warning if conditions change.

Finally, there was no need to use a grinder. The paint could have been removed by chipping or with a paint-removal solvent. When flammable materials are handled, it is good practice to add an extra layer of safety by not using sources of ignition when safer methods are practicable.

Ethylene oxide can be ignited and decompose, producing both heat and a rise in pressure, in the absence of oxygen. However, some oxygen will have been present in the incident described because the manway was open.

REFERENCES

1. Kletz, T.A. (1995). Equipment maintenance, in *Handbook of Highly Toxic Materials Handling and Management*, S.S. Grossel and D.A. Crowl eds., Dekker, New York, Chapter 11.

2. Woodward, J.L. and J.K. Thomas (2002). Lessons learned from an explosion in a large fractionator, *Proceedings of the AIChE Annual Loss Prevention Symposium*, March.

3. Anon. (2001). *Refinery Fire Incident*, Investigation Report No. 99-0141-CA, Chemical Safety Hazard Investigation Board, Washington, D.C.; summarized in *Loss Prevention Bulletin*, Oct. 2002, 167:4–7.

4. Gillard, T. (1998). Dangerous occurrence involving an instrument air compressor, *Loss Prevention Bulletin*, August, 142:16–18.

5. Anon. (2002). Corrosion dangers from redundant pipework. *Loss Prevention Bulletin*, Feb., 163:22.

6. Giles, D.S. and P.N. Lodal (2001). Case histories of pump explosions while running isolated. *Process Safety Progress*, **20**(2):152–156.

7. Bickerton, J. (2001). Near-miss during commissioning. *Loss Prevention Bulletin*, Dec., 162:7.

8. Anon. (2000). *Safety Alert*, Mary Kay O'Connor Process Safety Center, College Station, TX, 12 Sept.

9. Ogle, R.A. and R. Carpenter (2001). Lessons learned from fires, flash fires, and explosions involving hot work. *Process Safety Progress*, **20**(2):75–81.

10. Anon. (2002). *Investigation Report: Steel Manufacturing Incident*, Report No. 2001-02-1-N, Chemical Safety and Hazard Investigation Board, Washington, D.C.

Chapter 2

Entry into Confined Spaces

A woodcutter who spends most of the day sharpening his saw and only the last hour of the day cutting wood, has earned his day's wage.

— Menachem Mendel of Kotzk (1787–1859).

In the same way, time spent preparing equipment for entry is time well spent.

Many people have been killed inside tanks and other confined spaces. Sometimes they have entered without permission to do so or merely put their head inside an open manway to inspect the inside. Sometimes entry was authorized but not all of the hazardous material had been removed or it had leaked back in because isolation was poor. Sometimes hazardous material was deliberately introduced in order to carry out tests. Sometimes people have entered a confined space to rescue someone who has collapsed inside and been overcome themselves (see Sections 3.3.1 and 6.1.1).

2.1 INCOMPLETE ISOLATION

A trayed column was prepared for inspection. It was emptied, the remaining vapor removed by sweeping with nitrogen, and the nitrogen replaced by air. All the connecting lines were blinded — or so it was thought — and tests showed that no toxic or flammable vapor was present.

All this preparatory work was done on the night shift but the signing of the entry permit was left to a day superintendent. As he had been assured that all necessary precautions had been taken, he signed the permit. On the way back to his office he passed near the column, so he stopped to have a look at it. He heard a slight hissing noise and traced this to two instrument connections that had not been blinded. These instruments measured the pressure difference between the column and another column mounted on top of it. This column was still in use. The two instrument connections were insulated along with other lines and had been overlooked. The superintendent canceled the entry permit. Tests showed there was hydrocarbon in the column.

We can learn much from this near miss:

- When preparing a vessel for entry, give the maintenance team a list of all lines to be blinded (or disconnected), identify each one with a numbered tag and, if there is any doubt about their location, mark them on a sketch. Never ask them to blind *all* lines. (*similar* is another word that should never be used; see Sections 9.1 and 9.2.)
- Check that all the lines have been blinded (or disconnected).
- The person who signs the entry permit should always carry out his own check, regardless of any checks carried out by other people. His signature is on the permit and he is legally and morally responsible if anything goes wrong [1].

Although not relevant to this incident, note that if any lines are already blinded, the blinds should be removed and checked to make sure that they are not weakened or holed by corrosion.

2.2 HAZARDOUS MATERIALS INTRODUCED

2.2.1

A man was cleaning a small tank ($36\,m^3$, 9,500 gallons) by spraying the inside with cyclohexanone. He was killed by chemical exposure, lack of oxygen, or a combination of both. Two other men were killed while trying to rescue him. As well as being toxic, cyclohexanone is flammable [2]. No one should enter a confined space unless the concentration of flammable gas or vapor is <25% of the lower flammable limit. Air masks should be worn if the concentration of toxic vapor is above the threshold limit value or, for very short exposure, above the appropriate limit.

Entry should not normally be allowed even with air masks into atmospheres which are irrespirable, either because the oxygen content is too low or the concentration of toxic gases could cause death or injury in a short time. If such entry is permitted, two people trained in rescue and resuscitation should be on duty outside the vessel. They should have available all the equipment necessary for rescuing the person inside the vessel and they should always keep him or her in view (see *WWW*, Section 11.5).

2.2.2

Three men were overcome while cleaning a tank with trichloroethane. Never take hazardous liquids into a confined space unless the spillage of the total amount introduced will not cause the vapor concentration to exceed the threshold limit value or 25% of the lower flammable limit. Forced ventilation can be used to reduce the concentration of vapor but if so, the concentration should be monitored.

2.2.3

A 20-yr-old contract worker who was cleaning the inside of a paint-mixing tank took a bucket full of methlyethyl ketone, a flammable solvent, into the tank. When the vapor exploded the worker suffered 70% burns and died in a hospital. The source of ignition was static electricity generated when he repeatedly dipped his scouring pad into the bucket. In court the company said that they now used remote cleaning methods [3]. Why didn't they do so before? The accident was not hard to foresee in the light of previous experience. There is more on static electricity in Sections 3.2.7, 6.2.5, 8.1, and 10.7.

2.2.4

On other occasions, welding torches have been left inside confined spaces during a meal break or overnight. Welding gas has leaked, resulting in a fire or explosion when the torch was lit. Or argon has leaked and the welder has been asphyxiated on reentering the confined space.

2.3 WEAKNESSES IN PROTECTIVE EQUIPMENT

Compressed air supplies to air masks have failed for various reasons. For example:

- Hoses have been attached to connectors by crimped rings or by the type of fasteners used for water hoses in cars. These are not suitable for industrial use. Bolted connections are better [2].

- Air filters have been blocked by ice in cold weather.

- Emergency supplies of compressed air have failed, either because the emergency cylinders were empty or the change-over mechanism failed to operate. Emergency supplies should be tested each time the air mask is used.

- Nitrogen cylinders have been connected to compressed air lines in error. Different types of connections should be used for nitrogen and air. Many people have been overcome by nitrogen — another example follows — and the odorizing of nitrogen has therefore been suggested [4].

2.4 POOR ANALYSIS OF ATMOSPHERE

A nitrogen receiver, 8 m (26 ft) tall and 2 m (6.5 ft) wide was due for inspection. The inlet line was disconnected and the manway cover removed. The manway was near the bottom of the vessel and there was no opening at the top. The vessel was purged by natural ventilation supplemented by a compressed air hose and a test showed that that the oxygen content was normal.

An inspector entered the vessel and inspected it from a permanently fitted internal cat ladder. The standby man heard a noise, looked into the vessel, and saw that the inspector had fallen off the ladder. When the standby man tried to enter the vessel to rescue the inspector, he found that his self-contained air mask was too big to go through the manway. Fortunately, the emergency services arrived in a few minutes and were able to rescue the inspector, who had suffered more injury from his fall than from the low oxygen content.

Why did the analysis give a false reading? The test was carried out near the open manway at the bottom of the vessel. There was less oxygen near the top. Tests should always be carried out at various parts of a confined space (unless it is very small). This could have been done by removing a blank at the top of the vessel (this would also have improved the ventilation) or by using a long probe [5].

The need to test in more than one part of a vessel is hardly a new discovery (see Section 1.10.2). It has been known for many years and described in published reports. But it was unknown in the plant where the incident occurred (or if known it had been forgotten). Unfortunately, accident reports rarely tell us what training employees had received or what books, magazines, and safety reports were available for them to read.

Before anyone enters a confined space, we should ask how he would be rescued and by whom, if he collapsed, for any reason. We should make sure that the standby man is properly trained and equipped. The only good feature in this incident is that the standby man did not try to enter the vessel without an air mask. Many people have done so, to rescue someone overcome inside, and themselves been killed or injured (see also Sections 3.3.1 and 6.1.1).

2.5 WHEN DOES A SPACE BECOME CONFINED?

The inside of a storage tank or pressure vessel is obviously a confined space, and before anyone is allowed to enter it, a systematic procedure should be followed. The tank or vessel should be isolated by disconnecting or blinding of connecting lines; it should then be cleaned, the atmosphere tested, and air mask specified if necessary. However, some confined spaces are less obvious. If a tank is being built or a hole dug in the ground, when do they become confined spaces? A rule of thumb often used is to treat them as confined spaces when the depth is greater than the diameter. Leonardo da Vinci's advice on town planning over 500 years ago was, "Let the street be as wide as the height of the houses."

The following incidents show how easily people can unwittingly fail to recognize confined spaces.

2.5.1

Two men used liquid nitrogen to freeze water lines in a trench, as part of a cut-and-weld job. There was too little ventilation to disperse the nitrogen as it evaporated and they were overcome. No safety harnesses were worn and no oxygen meter was used.

2.5.2

During a plant shutdown, a piece of equipment was removed from a 1.2-m (48-in) diameter pipe. No one entered the pipe but the inside was inspected by shining a light into it. Bright sunshine made it difficult to see anything so a black plastic sheet was draped over the end of the pipe. There was a strong breeze so to hold the sheet in place two men sat on one edge of the sheet and two others held it over them. The two sitting men then inspected one of the open ends.

They then tried to do the same at the other open end of the pipe. Unfortunately, there was a flow of nitrogen coming out of the end of the pipe and the two men were overcome. One died and the other recovered after five days in hospital.

Both the man who died and his coworker were men of great experience. The day before, one of them had asked for nitrogen to be injected in order to protect the catalyst. The injection point was $\approx 50\,m$ (150 ft) and several floors away and he may not have realized that the nitrogen would exit through the 48-in pipe. He certainly did not realize that a plastic sheet held loosely over the end of the pipe turned it into a confined space [6].

The company's entry procedure did not draw attention to the hazards of temporary enclosures. Obviously it should have, but even if it had, would the men have remembered this fact? Instructions are no substitute for knowledge and understanding, that is, knowledge that confined spaces can easily be formed; knowledge that nitrogen in quite small amounts can reduce the oxygen level to a dangerous extent; and knowledge that what goes in must come out and that whenever we put anything into a plant we should ask where it or something else will exit. The root cause of the accident was the failure of the company to give their employees this understanding of the hazards (see also Sections 7.4, 7.5, 8.12, and 14.5).

2.5.3

To save energy, a company decided to use a flammable and toxic waste gas (known as tail gas) to run a diesel engine and generate power. The gas first had to be cooled and this was carried out in the equipment shown in Figure 2-1. Two pumps were located inside the skirt of the column. One pumped some of the wash water from the base of the column into the venturi and the other circulated the bulk of the water through a cooling tower and back to the top of the column. There were four arched openings in the base of the column so it was not considered a confined space. However, the location was very congested and this reduced ventilation.

An electrician and an engineering student were asked to repair the circulation pump. The procedure they were told to follow, never written

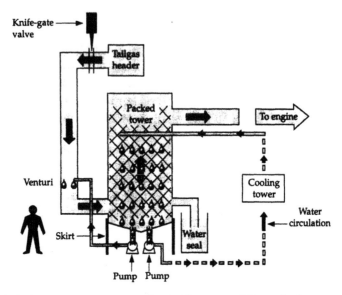

Figure 2-1. A man was overcome by leaking gas while replacing a pump in a confined space. From reference 7. Reprinted with the permission of the Institution of Chemical Engineers.

down, was to close the knife-gate valve, thus stopping the flow of gas to the column, and then to get rid of the gas already in the column as well as any leaking past the valve — a type that does not give complete isolation — by draining the water seal. There were no valves in the suction lines to the two pumps.

Unfortunately, the electrician and the student forgot to drain the water seal. Nevertheless, they removed the pump and blanked the open end without incident. They then refilled the tower with gas to prevent air from leaking in. Later that day they replaced the pump, presumably using the same procedure as before. While they were working on their hands and knees, they felt unwell. Before they could get out, the student noticed that the electrician had become unresponsive. Fortunately, the student was able to pull him out and he soon recovered.

One of the causes of the incident was the failure to recognize that the space inside the skirt was a confined space. There was also much else wrong:

- It is not good practice to locate a pump (or any equipment that needs regular maintenance) inside a column skirt (or anywhere else where access is poor). Maintenance will be better as well as safer when access is good.

- Pumps are normally fitted with suction and delivery valves, which are closed before the pumps are removed or repaired. The method followed on the plant: removing a pump (or any other equipment) without isolating it by closed valves and then trying to blank the open ends of the connecting pipes before much of the contents leak out or air leaks in, is not good practice.

- All instructions except the simplest should be written down, not just passed on verbally. Many tasks are part of the skill of the craft and it is not necessary to tell skilled craftsmen how to carry them out; however, that does not apply to detailed and unusual procedures.

- Normally, a member of the operating team prepares equipment for maintenance and then issues a permit-to-work which is accepted by the senior member of the maintenance team. The involvement of two people with different functions and the filling-in of a permit provides an opportunity to check that all necessary precautions have been taken. This opportunity is lost when the same person prepares the equipment and then carries out the repairs. In such cases, it is good practice for such a person to complete a check list, in effect, issuing a permit to him or herself, or for a colleague to do so for him.

Underlying these detailed causes were managerial failures. The project was being carried out by a special team whose members undertook their own maintenance, independently of the normal operating and maintenance organizations. Research and development workers often believe that they can carry out work on plants without being confined to the normal safety procedures. It should be made clear to them that they cannot. Also, to quote from the report [7], "the deadlines were seen as very challenging by those involved," a euphemism suggesting that speed was put before safety.

2.6 MY FIRST ENTRY AND A GASHOLDER EXPLOSION

After I graduated, I spent the next seven years in the Research Department of Imperial Chemical Industries (ICI) at Billingham, UK. After two years I was sent down to the factory for about six weeks to see how the company earned its profits. One quiet Saturday afternoon one of the shift foremen asked me if I would like to go inside a gasholder. It was the dry type in which a moveable disc separates the gas in the lower part of a cylinder from the air in the upper part and there is a tar and canvas seal between the disc and the cylinder walls.

At a guess the volume of the gasholder was several thousand m^3 and its height was $\approx$ 3$\times$ its diameter. To get inside it, we had to go up a staircase onto the roof, through an opening in the center, and then down a Jacob's ladder, a folding ladder, onto the disc. As Figure 2-2 shows, during half the descent we were clinging to the wrong side of the sloping ladder. Fortunately, the gasholder was nearly empty and the angle of the ladder was not too great.

I cannot remember what was in the gasholder. It may have been coke oven gas, water gas (H_2 + CO) or producer gas (N_2 + CO_2). The atmos-

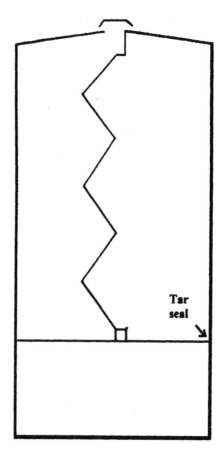

Figure 2-2. Diagram of dry gasholder showing Jacob's ladder.

phere above the disc in the gasholder was not tested before we entered and there was no entry permit and no standby man, although the foreman mentioned that strictly speaking there should have been one (this was many years ago). The whole experience was eerie and I have never forgotten it.

The foreman mentioned that following an explosion in Germany, dry gasholders were out of favor, ICI would not build others, and the long-term plan was to replace them with wet gasholders. I never found out what had happened in Germany but I found an article on a dry gasholder explosion there, which is worth recounting [8].

The gasholder contained coke oven gas. A section of the bypass pipe was removed for cleaning as it was partially blocked with naphthalene. On the inlet side, the section of pipe was isolated by a closed valve (Figure 2-3) and on the outlet side by a blind slip-plate. When the missing section was replaced, it was found that the pipe coming from the valve had sunk and the two pipes could not be lined up. It was then decided to remove the support at the end of the replaced section so that it would

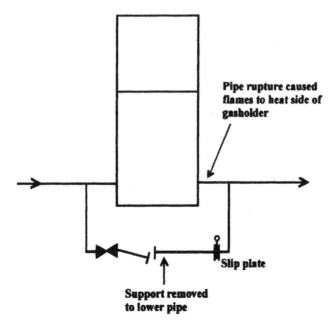

Figure 2-3. Gas leaked through closed valve while burning was in progress nearby.

also sink. This involved welding. It ignited gas which had leaked through the closed valve and the resultant explosion tore the outlet main close to the gasholder. The flame from this much larger leak went up the side of the gasholder and five minutes later the gasholder exploded. Either the heat distorted the walls or evaporated the tar. Either way, this would allow gas to bypass the disc and mix with the air in the upper portion of the gasholder. (See Section 12.3 for an account of another leak through closed valves.)

Perhaps this was the explosion in Germany to which the ICI foreman referred. However, he said that the German explosion occurred because the disc in a gasholder tilted and jammed and gas got past it, and I recall seeing a tilt alarm on the disc in the ICI gasholder.

2.7 FAILURE OF A COMPLEX PROCEDURE

Confined spaces are usually entered through a manway or similar opening which has to be unbolted before anyone can enter. This makes unauthorized entry difficult, though not impossible, because someone has to unbolt the entrance. At one plant there was a room containing hazardous materials and equipment which was treated as a confined space. Entry was through a door, which was normally locked shut. To make sure that no one could be accidentally locked inside the room, the following rather complex procedure was followed:

- The key to the room was kept in a box fitted with several locks, each operated by a different key. The process foreman kept all the keys. Normally only one lock was closed.

- If someone needed to enter the room, the foreman first established that it was safe to do so. He then issued a permit-to-work and gave the box key to the person who was going to enter. He or she then opened the box, got out the door key, opened the door, LOCKED THE DOOR KEY BACK IN THE BOX, and kept the box key.

- If more than one person was entering the room, each of them was given one of the box keys and they each locked the box with the door key inside. The box had a window so that it was possible to see whether or not the door key was in it. It was therefore impossible to lock the door of the room until everyone inside had left and unlocked the box.

Before you read on, please make sure you understand the way the system works, because some of the people at the plant did not.

This system seems good, though complex, but like all systems it could and did degrade. One day a mechanic had to enter the room for a quick job so he left the door key in the door. (Afterwards the foreman admitted that this was often done for quick jobs.) Before the mechanic had finished, two other men arrived to erect scaffolding for a later job. As the door was open, they did not bother to get box keys from the foreman. When the mechanic finished his job, he left the door key in the door so that the scaffolders could lock the door when they finished. He did not remind them to do this; he assumed they knew what to do.

Before the scaffolders had finished, another mechanic arrived to carry out the second job. He went first, as usual, to the foreman's office to get a permit-to-work and a box key. The foreman was not there but the fitter saw the permit for the first job on the table. He added his job to the permit and signed it. He took another box key (and one for his coworker) and both of them locked the box but they did not notice that the door key was not in the box.

Later someone noticed that the door was open but the key was in the door.

2.7.1 What Went Wrong?

- It is bad practice to allow people to add extra work to an existing permit-to-work — it had become custom and practice at the plant — and they should certainly never do so without the signed agreement of the person who issued the permit.

- If more senior people had kept their eyes open, they might have spotted sooner that procedures were not always followed. Fortunately, this incident drew it to their attention before an accident happened. It shows the value of following up dangerous occurrences.

- Many of the people at the plant did not fully understand the procedure or the reasons for it.

- Could a simpler procedure be devised? For example, each person padlocking the door open with his own padlocks when an entry is in force. Simpler procedures reduce the temptation to take shortcuts.

- If simplified procedures are allowed for quick jobs, or become custom and practice, how do you deal with quick jobs that become long ones?

(Every do-it-yourself enthusiast knows that 5-min jobs can take all day.)

- Several months earlier, two men had followed the correct procedure before entering the room. But they had to enter again soon afterwards and this time they left the key in the door while they were in the room. Subsequent inquiries showed that they did not understand the reasons for the procedure and were just following it blindly. No wonder they could not be bothered to go through it a second time. If this incident had been followed up more thoroughly, the second incident might not have occurred.

Section 14.6 describes an overly complex instrumented system for controlling entry which also failed.

2.8 AVOIDING THE NEED FOR ENTRY

When planning an entry, the first question to ask should be, "Can we avoid the need?" The following are common reasons for entry and possible ways of avoiding the need:

- TO INSPECT OR REPAIR EQUIPMENT INSIDE THE VESSEL: withdraw the equipment from the vessel.

- TO INSPECT THE INSIDE OF THE VESSEL: if doctors can inspect the insides of our stomachs, bladders, and bowels from outside (and display the insides on a screen while doing so), engineers should be able to do the same with vessels.

- TO CLEAN THE VESSEL: use high-pressure water combined with a large bottom opening if solids have to be removed. In the early days of polyvinylchloride, production men entered and cleaned the reactors after every batch. When frequent exposure to high concentrations of vinyl chloride was found to cause cancer, other methods of cleaning were developed.

- TO OPERATE OR MAINTAIN VALVES ON VESSELS IN PITS: do not put vessels in pits but, if you have already done so, consider remote operation of valves. If you insist on putting vessels in pits, provide ample room between the vessel and the walls of the pit.

- TO CLEAR BLOCKAGES IN SILOS: use a remotely operated portable flail [9].

• TO CONSTRUCT THE VESSEL: could low-pressure vessels be constructed from the outside? In the construction of some UK railway carriages, components are fixed to the floor, roof, and the two walls before these four pieces of steel are bolted together. Access is much easier when everything can be at a convenient height [10].

REFERENCES

1. Gillard, T. (1998). Entry into vessels — a near miss. *Loss Prevention Bulletin*, Oct., 143:18.

2. Donaldson, T. (2000). Confined space incidents — a review. *Loss Prevention Bulletin*, Aug., 154:3–6.

3. Anon. (2000). DuPont in dock over fireball death. *Loss Prevention Bulletin*, Aug., 154:27.

4. Turner, S. (1988). Odorise your nitrogen. *The Chemical Engineer*, 9 July, 661:20–21.

5. Anon. (2000). Vessel inspector overcome by nitrogen. *Loss Prevention Bulletin*, Aug., 154:20.

6. Hill, P.l., G.P. Poje, A.K. Taylor, and I. Rosenthal (2000). Nitrogen asphyxiation. *Loss Prevention Bulletin*, Aug., 154:9.

7. Porter, S.R. and P.J. Mullins (2000). Waste gas leads to near fatality. *Loss Prevention Bulletin*, Aug., 154:15.

8. Anon. (1933). The disaster at the Neunkirchen Iron Works, *The Gas World*, 22 April, p. 397 (Translation from *Das Gas und Wasserfach*, No. 14, April 8 1933).

9. Health and Safety Executive. (1999). *A Recipe for Safety*, HSE Books, Sudbury, UK, p. 18.

10. Abbott, J. (1999). Turbostar comes of age. *Modern Railways*, Dec., **56**(615):891.

Chapter 3

Changes to Processes and Plants

Midas, a legendary king of Phrygia, asked the gods to make everything he touched turn into gold. His request was granted but as his food turned into gold the moment he touched it, he had to ask the gods to take back their favor.

Unfortunately, the gods are less obliging today and will not reverse the results of ill-considered modifications.

It is now many years since the explosion at Flixborough in 1974 (see *WWW*, Section 2.4) brought home to the process industries the need to look systematically for possible consequences before making any change to plants or processes. Many publications [1,2,3] have described accidents that occurred because no one foresaw the results of such changes and suggested procedures for preventing such accidents in the future. Nevertheless, as the following examples show, unforeseen consequences still occur. Sometimes there is no systematic procedure, sometimes the procedure is not thorough or is not followed, and sometimes the change is so simple that a formal review seems unnecessary. There is also a reluctance in many companies to look in the literature for reports of similar situations. According to an experienced process safety engineer:

People make very little preparation for a management of change or process hazards analysis (PHA) by looking at the literature or making a search for events in similar facilities. We can sometimes prompt them to look at events within their own facility . . . but getting them to spend any reasonable time reviewing other events is tantamount to pulling teeth. . . . I would estimate that less than one in twenty PHA practitioners expend more than a very small effort in such preparations [4].

Chapter 2 of *WWW* described mainly changes to equipment — following Flixborough that seemed the main problem — so to restore balance this chapter describes more changes to processes. The next chapter describes changes to organizations.

3.1 CHANGES TO PROCESSES

3.1.1 Scale-up is a Modification

In scaling up a process from the laboratory to production-scale, a company changed it from semi-batch operation, in which the second reactant is added gradually, to batch operation, in which the entire quantities of both reactants are added at the outset. Twenty percent of the batches showed temperature excursions but the operators were able to bring them under control by manual operation of cooling water and steam valves. The company then increased the reactor size from 5 to $10\,m^3$ (1,350 to 2,700 gal) and increased the quantities of reactants by 9%. The proportion of batches showing temperature excursions rose to half and ultimately the operators failed to keep a batch under control. The manhole cover was blown off the reactor and the ejected material caught fire. Nine people were injured.

The company failed to use its management-of-change procedure and also failed to respond to the rising number of temperature excursions [5].

Failures to understand scale-up go back a long way. Canned food was introduced in 1812. In 1845 it became part of regular British Royal Navy rations. Some time later there was an outbreak of food poisoning. Larger cans had been used and the heat penetration became insufficient to kill the bacteria in the middle [6].

3.1.2 Unrecognized Scale-up

In his biography, *Homage to Gaia* [7], James Lovelock describes an incident that occurred when he was working for a firm of consultant chemists. There had been a sudden deterioration in the quality of the gelatine used for photographic film, and he and another chemist were sent to visit the manufacturers. They asked the foreman if anything had changed. He replied that nothing had changed; everything was exactly as before. Lovelock's colleague noticed a rusty bucket next to one of the vessels and asked what it was for. The foreman said that a bucketful of hydrogen peroxide was added to each batch of gelatine but as the bucket was rusty he had bought a new one the previous week. "We soon solved the firm's problem when we found that the new bucket was twice the volume of the old one." Its linear dimensions were only 25% greater but the foreman had not realized that this doubled the volume.

3.1.3 Ignorance of a Reaction

Lovelock also described a modification that nearly took place but was prevented in time. The United Kingdom Gas Board, at the time the monopoly supplier of natural gas, decided to label the gas in one of their major high-pressure gas pipes with sulfur hexafluoride to detect leaks along the pipeline. The technique would have worked well. Unfortunately, they did not know that a mixture of methane and sulfur hexafluoride will explode almost as violently as a mixture of methane and oxygen. Fortunately, they found out in time and abandoned their plan [8].

3.1.4 Changes Made to Handle Abnormal Situations

A coker is a large vessel, typically $\approx 12\,m$ (40 ft) tall, in which hot tar-like oil, after being heated in a furnace, is converted to lighter oils, such as gasoline and fuel oil, leaving a tarry mass in the vessel. On cooling, usually with steam and then water, this forms coke, which is dug out. A power failure occurred when a coker was 7% full and the plant was without steam for 10 h. The inlet pipe became plugged with solid tar and the operators were unable to inject steam.

There were no instructions for dealing with this problem, although a somewhat similar one had occurred two years earlier. On that occasion it had been possible to inject water to cool the contents but nevertheless when the bottom cover was removed from the coker, a torrent of water, oil, and coke had spewed out. When the second incident occurred, the supervisor therefore decided to let the coker cool naturally before opening it. Two days later the temperature of the outside of the bottom flange of the coker had fallen from its usual value of 425 °C (800 °F) to 120 °C (250 °F) so the supervisor decided to go ahead. The operators injected some steam — presumably through a different route than the normal one — to remove volatile products and then started to open the coker. The top cover was removed without incident. The bottom cover was unbolted while supported as usual by a hydraulic jack. When the jack was lowered, hot vapor and oil gushed out and immediately ignited. It was probably above its auto-ignition temperature. Six people, including the supervisor, were killed.

The immediate cause was a failure to realize that the temperature of the middle of the vessel was far higher than that of the walls, high enough to continue to convert the tar to gasoline. Afterwards, calculations showed that it would take two weeks, not two days, for the temperature to fall to a level at which it would be safe to open the coker. (Section 8.7 and 13.7 describe the results of other failures to calculate effects.)

The controls for the hydraulic jack should have been located farther away from the coker and so many people should not have been allowed so near.

One underlying cause was the failure to plan in advance for a loss of power. Plans should have been made for this foreseeable event but never were, even though an event had occurred two years before and caused a serious spillage,

Another underlying cause was the lack of technical support. The supervisor seems not to have been a professional engineer or recognized the need to consult one. The report [9] does not say whether or not there had been any downsizing or reduction in support but the incident is rather similar to that at Longford (see Section 4.2) where the operating team was also unaware of a well-known fact, in this case that metal becomes brittle when cold.

We have all been given, at some time, a food such as pasta or rice pudding, straight from the oven in the dish in which it was cooked. If it is too hot to eat, experience tells us that the outside bits are cooler and we eat them first. We know the outside cools faster than the inside. Unfortunately, we find it difficult to apply in one situation the lessons we have learned in another; they are kept in different parts of our minds.

3.1.5 An Abnormal Situation Produced by a Process Change

Powdered aluminum chloride, a catalyst, was added to a reaction mixture. A change was made: aluminum powder was used instead as it was expected to form aluminum chloride by reacting with the hydrogen chloride already there. Unfortunately, the reactor became choked with sludge. The aluminum was much denser than the aluminum chloride and the agitator was unable to prevent it from settling. If there was a management of change procedure — the report [9] does not say — no one considered the results of more or less mixing, an obvious question to ask when a hazard and operability study (Hazop) is carried out on a vessel in which mixing takes place.

The problem now was how to get the sludge out of the reactor. A chemist examined a sample. It reacted with water, producing heat, so he recommended that a large amount of water, eight times the weight of the sludge, should be put into the reactor as rapidly as possible. Someone suggested that a short burst of steam should first be put into the reactor to break up the sludge. The day supervisor agreed and gave his instructions by telephone to the afternoon shift supervisor who told the night shift supervisor who told an operator. By this time the instruction had become distorted and steam was added continuously for several minutes. The reactor exploded. Fortunately, no one was seriously injured.

An immediate cause was the sloppy method of passing on instructions despite the fact that the addition of water was known to generate heat. The instructions should have been precise and in writing and should have specified the duration of the steam burst. *Short*, like *all* and *similar* (see Section 9.1), is an imprecise word and should never be used in plant instructions.

Another immediate cause was the failure to calculate what temperature would be reached by the addition of steam and then water and the amount of gas that would be driven off. In the laboratory test, it dispersed easily. At the plant, the size of the vent was quite inadequate.

Underlying these causes was the assumption, as in the last item, that the supervisor can improvise changes in procedure to cope with an abnormal situation. In emergencies, he or she may have to do so but when possible these situations should be foreseen and planned for in advance. In the case just described, a few hours', or even a day's, delay while the proposed change was discussed by a group of people, including professional staff, and approved at an appropriate level, would have mattered little. Blowing up the reactor caused rather more delay, and cost $13 million.

3.2 CHANGES TO PLANT EQUIPMENT

3.2.1 Changes in the Direction of Flow

Figure 3-1a shows the original design of a reactor. Hot feed gas (PF) was passed upwards through catalyst tubes 3 m (10 ft) long. The heat of reaction was removed by circulating a molten salt through the shell. Note that the flow of this liquid was also upwards.

The hot exit gases were cooled in a waste heat boiler. During a review of the model, the contractors pointed out that the boiler required an expensive structure to support it. They could avoid this cost, they said, by reversing both flows through the reactor, as shown in Figure 3-1b and putting the boiler at ground level. This was agreed.

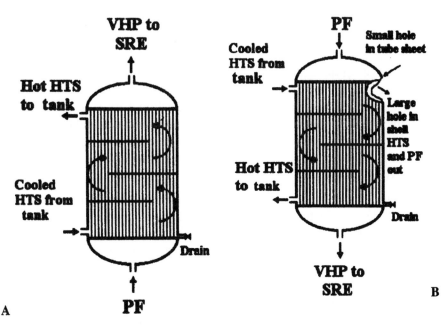

Figure 3-1a,b. (a) Original design of reactor; (b) Reversing the flow allowed a gas bubble to be trapped below the top tubesheet and led to overheating of the tops of the tubes. PF = Preheated feed; HTS = Heat transfer salt; VHP = Very hot product gases; and SRE = Steam raising exchanger. From reference 10. Reprinted with the permission of the Institution of Chemical Engineers.

Soon after start-up, some temperatures were erratic, violent vibrations occurred, and then the shell ruptured at a point opposite the liquid inlet line.

The investigation [10] showed that gas bubbles had been trapped under the top tubesheet. The tubes near it overheated and the nitrate coolant reacted with the iron shell in a thermite type reaction, the iron replacing the positive ions in the salt. The investigation also showed that gas bubbles can collect under a horizontal surface when the flow is downwards but not when it is upwards. The plant was rebuilt to the original design.

It might have been possible to retain the downward flow and to vent any gas bubbles back to the suction tank in the salt system. However, preventing the formation of bubbles is better than letting them form and then getting rid of them.

The change in design was made without following the company's normal procedure for control of change. Reversing the flow seemed such a minor change (and one that would save so much money) that it received no systematic appraisal.

Another incident was the result of an even more minor change in flow [11]. Four 16-m^3 (4,200-gal) tanks that held an odorizing liquid had to be taken out of service and cleaned. Three tanks were cleaned without incident. The procedure was to empty each tank, wash it with methanol, and then remove the last traces of the odorizing liquid by washing with a sodium hypochlorite solution. The tanks were normally blanketed with natural gas and this was left in use during the cleaning operation, the natural gas flowing through the ullage space of the tanks and then to the flare sytem.

When the fourth tank was cleaned, the arrangement of the pipework made it impossible for the natural gas to flow through the tank so the natural gas was just connected to it in order to maintain the pressure in the tank. During the hypochlorite wash, there was an explosion in the tank and flames were discharged through the relief valve. Tests then showed that there was 80% oxygen in the ullage space of the tank. The oxygen was probably formed by decomposition of the hypochlorite, catalyzed by the nickel in the stainless steel of the tank. While the first three tanks were cleaned, the continuous flow of natural gas swept out the oxygen as fast as it was formed.

Natural gas or other fuel gases are often used for blanketing when nitrogen is not available. They are just as effective as nitrogen in maintaining a pressure in the equipment and preventing air leaking in; however, if air does leak in, a fire or explosion is more likely to occur.

3.2.2 Two Changes in Fire-Fighting

Two buildings were 23 m (75 ft) apart. The same fixed fire-fighting unit served both buildings, as they were too far apart for a fire in one to spread to the other. Eight years later, a new building was built between the two original ones and one of the originals was demolished. The gap between the two buildings was now only 4 m (12.5 ft). When a fire occurred in one building, it spread to the other and the fixed equipment was too small to control both fires [12].

The heating system in a building had to be shut down for repair over a weekend. There were fears that the water in the sprinkler system might freeze so it was replaced by ethanol. You can guess what happened.

3.2.3 Adding Insulation is a Modification

To save energy, a company decided to insulate a valve, shown in Figure 3-2, which operated at 310 °C (600 °F). The three long bolts expanded and

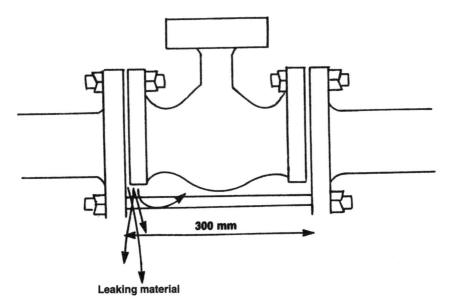

300 mm

Leaking material

Figure 3-2. When this valve was insulated to save heat, the long bolts expanded and the flanges leaked.

a leak occurred and ignited. The flames were 12 m (40 ft) long. The valve body, in direct contact with the hot liquid, would hardly have been affected by the insulation but the long bolts rose in temperature. A rise by 250 °C (450 °F) would increase their length by 1 mm (0.04 in) [13].

When the company decided to use valves with long bolts, they should have considered this as a modification and looked for possible consequences. These valves are not suitable for equipment which undergoes changes in temperature.

Another company was more successful. They went through their modification procedures when they reviewed a proposal to fit acoustic insulation to some pipework. They then realized that the acoustic insulation would also act as thermal insulation and prevent the cold gas in the pipes from picking up heat from the atmosphere. The insulation was still fitted but other changes were made to handle the change in temperature.

3.2.4 Two Unauthorized Changes

Figure 3-3 shows a 3-way cock in which the top of the central bolt has been marked to show the position of the cock. This was presumably done because the marks on the plug itself are hard to see. They are just faintly visible in the photograph.

Originally the marks on the bolt corresponded to those on the plug. At some time two washers were placed underneath the bolt. It could no longer be screwed right in and the marks no longer corresponded.

Most of the operators on the unit set the cock according to the marks on the bolt. Ultimately, this led to misdirection of a process stream, formation of an explosive byproduct, and an explosion.

3.2.5 A Very Simple Change

A company decided to display hot-work permits on the job. They were fixed to any convenient item of equipment. On one unit they were pushed into the open end of a 1.5-in pipe. The man who did this probably thought it was a scaffold pole or a disused pipe. The pipe actually supplied a controlled air bleed into a vacuum system to control or break the vacuum. The hot-work permits were sucked into the pipe and blocked the motor valve in the pipe. Product was sucked into a condenser and the unit had to be shut down for cleaning for two days. Several permits were removed from the valve.

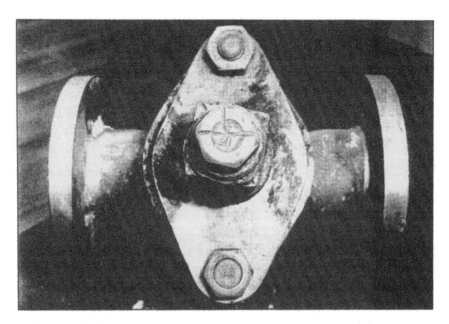

Figure 3-3. The central bolt was marked to show the position of the cock. When two washers were inserted under the bolt, it could not be screwed down as far as before and the marks no longer corresponded with the position of the cock.

3.2.6 A Temporary Change

A drum-filling machine was causing problems during a Friday night shift. The shift foreman decided to change over to manual filling until the maintenance team returned to work on the following Monday. Until then the valve on the filling machine had to be opened with an adjustable wrench. In addition, the filling head could not be lowered into the drum and the drum had to be carefully positioned under the filling head.

The maintenance team was too busy to attend to the filling machine and a week later the temporary system was still in use when the inevitable happened. A drum was not positioned accurately and the liquid hit the top of the drum, splashing the operator's face.

The report blamed poor communication [14]. The shift foreman's note in his log and in the job list did not draw attention to the fact that the temporary work method was hazardous and so the job got the priority given to an inconvenience, not a hazard. However, this is not very convincing. The unit manager, the other shift foremen, the fillers, and the safety representative, if there was one, should have spoken to the maintenance team and drawn attention to the hazard. In a well-run organization, written messages are for confirmation, precision, and recording; things get done by talking to the people who will have to do the work, asking them, persuading them, sweethearting them, call it what you will [15].

It is sometimes necessary to make changes at short notice in order to keep a plant running. In such cases, the normal procedures for the control of modifications should be carried out as soon as possible and no later than the next working day.

3.2.7 Another Trivial Change

Filters removed dust from a ventilation system. The dust fell into a 55-gal drum. From time to time, the drum was removed by a forklift truck, emptied by vacuum, and replaced. At some point in the life of the plant, the operators found that it was easier to replace the drum and position it correctly if they kept it on a wheeled trolley. They did not realize that, as the wheels had rubber tires, the drum was now an ungrounded conductor — and could accumulate a charge of static electricity, either during the vacuuming operation, during transport, or as a result of dust falling into it. The trolley was in use for a considerable time before conditions were just right for an explosion. While the trolley was being replaced, a charge passed from the drum to grounded metal nearby, igniting a small cloud of dust that fell into the drum from the filters at just that moment. As so often happens, the small initial explosion disturbed dust that had settled and was followed by a larger and more damaging explosion.

Static discharges may have occurred before but they happened at times when no dust cloud was present. On the day of the explosion, the atmospheric temperature was very high ($\approx 38\,°C$, $100\,°F$) and this would have lowered the ignition energy of the dust and made an explosion more likely [16].

This incident shows the limitations of instructions and the need to give operators an understanding of the hazards of the materials and equipment used. However many instructions we write, we can never cover every possibility. If we try to do so, our instructions get longer and more complex and fewer people read them. It is better to educate people so that they understand the hazards.

Auditors should look at plant instructions. Sometimes they are out-of-date or cannot be found. More often they are spotlessly clean, like poetry books in public libraries, showing that they are rarely consulted. (There is more on static electricity in Sections 2.2.3, 6.2.5, 8.1, and 10.7.)

3.2.8 Unintended Changes

These occur when suppliers supply the wrong process materials (see Section 9.3) or construction materials. For example, a chlorine vaporizer was shut down for repair and inspection. Soon after it was started up, a spiral-wound gasket, changed at the shutdown, blew out. The metal winding in the gasket could not be found. Trace metal analysis showed that it had been made from titanium, which reacts rapidly with chlorine, instead of nickel.

If use of the wrong process material or material of construction can have serious effects on safety, then all incoming materials should be tested before acceptance or use. This became commonplace in the 1970s, after a number of serious incidents, but many companies abandoned their checking programs when their suppliers obtained quality certification. How many more incidents do we need before they are reintroduced?

3.2.9 A Change to the Type of Valve

In carbon dioxide absorption plants, the gas is absorbed in potassium hydroxide, which becomes potassium carbonate. Control valves let down the potassium carbonate solution from high to low pressure. One plant used a motorized ball valve instead. When the jet from this type of valve impinges on a surface, it produces a ring-shaped corrosion groove. A disc of metal was blown out of a bend downstream of the valve.

Sensing the loss of pressure, the automatic controller opened the ball valve fully, discharging hot potassium carbonate solution out of the hole. Unfortunately, the pipe was opposite the control room window. The window was broken and all the operators were killed. It was shift change time and more operators than usual were present [17].

3.2.10 A Change in the Cooling Agent

A reactor was cooled by circulating brine through the jacket. The brine system was shut down for repair so town water was connected to the jacket. The gauge pressure of the town water (9 bar or 130 psi) was greater than the design pressure of the jacket's inner wall, which gave way.

The works modification approval form, which had been completed by the supervisor and the maintenance engineer, asked 20 questions, one of which was, "Does the proposal introduce or alter any potential cause of over/underpressurizing the system or part of it?" They had answered **No** [18].

3.2.11 A Failure to Recognize the Need for Consequential Change

When one item in a plant is changed, others may have to be changed to match, but this is not always recognized. A company filled drums with liquid chlorine. One was overfilled and bulged when the temperature rose so for protection a high-weight alarm was fitted to the filling and weighing machine. It was set at 1,400 kg. A change was made to smaller drums that were completely full at a gross weight of 1,335–1,350 kg but no one remembered to change the alarm setting. Either there was no procedure for the management of change or the change was considered so insignificant that the procedure was not followed. About three years later, another drum was overfilled and bulged. The company then decided to check weigh the drums and ordered an additional weighing machine.

Another three years later, this machine had arrived but had not yet been installed and a third drum was overfilled and bulged, this time at a customer's premises. The cause was a minor one-time change in the filling procedure. As the storage space was full, a drum was left connected to the filling machine overnight and the drum-filling valve was leaking or was not fully closed. Check weighing would have prevented this incident [19].

Fortunately, none of the overpressured drums burst or leaked though they were taken well above their design pressures. The large difference between the design pressure and the rupture pressure is a good example of defense in depth. Most pressure vessels can withstand several times their design pressure before they rupture but not all equipment is as strong (e.g., low-pressure storage tanks are quite fragile). In contrast, most equipment can withstand only a small percentage increase in absolute temperature. The life of furnace tubes is shortened if they are exposed to a few percent increase in absolute temperature for a short time.

3.2.12 An Example from the Railways

In the early days of railways, the gaps between the rails caused almost intolerable vibration. To reduce it some railway companies cut the ends of the rails diagonally so that they overlapped and formed a smoother joint. Unfortunately the spikes holding down the rails sometimes failed to do so, the end of the rail rose, the wheel went underneath it, and the pointed end of the rail went though the floor of the carriage. The person sitting above it was likely to be speared and impaled against the roof [20].

3.2.13 Another Historic Incident

Malaria and yellow fever, both spread by mosquitoes, hindered the building of the Panama Canal. The cause? To prevent ants climbing up the legs of hospital beds, they were set in pans of water — which unfortunately created an ideal breeding ground for mosquitoes [21].

3.3 GRADUAL CHANGES

If a frog is put into hot water, it jumps out. If it is put into cold water and the temperature is gradually raised, it stays there until it dies. In a rather similar way, we often fail to notice gradual changes until they have gone so far that an accident occurs. Section 2.9 in *WWW* describes several examples, including a gradual reduction in the flow through a steam main as the result of recession in the industry. The steam traps were barely adequate; this did not matter when the flow was large but when it became lower, condensate accumulated and water ruptured the pipe.

3.3.1 A Gradual Change in Concentration

Natural gas liquids were dried by passing them through a bed of molecular sieves, which also absorbed some hydrogen sulfide; the sieves were then regenerated by a stream of hot gas. They had to be changed every three or four years. The old ones were wetted with a fire hose in case any pyrophoric materials were present and to keep down dust and then poured down a chute into a high-sided tipper truck for disposal.

The sieves formed a mound shape and had to be spread level in the truck. A man who entered the truck to spread them collapsed. Three other men entered the truck to rescue him. All three collapsed; two of them and the first man died, poisoned by hydrogen sulfide. The sieves had a greater affinity for water than for hydrogen sulfide and released the gas when wetted.

There was much wrong. The high-sided truck was not recognized as a confined space and so the entry procedure (see Chapter 2) was not followed; the men filling the truck had not been warned that toxic gas might be present; many of the operators and staff did not know that it could be released; and no hydrogen sulfide detectors were supplied. But underlying all these shortcomings was the fact that over the years the amount of hydrogen sulfide in the natural gas liquids had gradually increased without anyone realizing that it had reached a level where change in procedures was necessary.

3.3.2 A Gradual Change in Maintenance

Another incident occurred in the UK railway system. This heretofore single organization was split into many independent private companies in the hope that this would provide competition and reduce costs. (It did not.) One company owned the track, other companies maintained it, yet more companies owned the trains, and a fourth set of companies maintained them. In an attempt to reduce cost, there was a gradual tendency to reduce maintenance but still maintain specifications. However, "Both sides of the wheel/rail interface may be operating within their respective safety-based standards, but the combined effect of barely acceptable wheel on barely acceptable rails is unacceptable" [22]. This led to rolling contact fatigue of the track (often called gauge corner cracking), a train crash at Hatfield near London in October 2000 that killed four people, and a consequent upheaval while hundreds of miles of rail were replaced.

The engineering principle involved is hardly new. In 1880, Chaplin showed that a chain can fail if its strength is at its lower limit and the load is at its upper limit [23] (see Section 15.5). The Hatfield crash did not occur because engineers had forgotten this but because there were no engineers in the senior management of the company that owned the track. They had all been moved to the maintenance companies. This accident is therefore also an example of the need for the management of organizational change, which is discussed in the next chapter.

3.3.3 Gradual Changes in Procedures

Gradual procedural changes are more frequent that gradual changes in equipment or process conditions. Procedures corrode more rapidly than steel and can disappear once managers lose interest. A procedure is relaxed, perhaps for a good reason. Perhaps there are technical reasons why the normal procedure for isolating equipment for maintenance cannot be followed on a particular piece of equipment. Nothing happens and the simpler procedure is used again just to save time or effort. Before long it has become standard and newcomers are told, "That instruction is out-of-date. We don't do it that way any more." To prevent this sort of gradual change, supervisors and managers should keep their eyes open and also explain why certain procedures are necessary. An effective way of doing this is to describe or, better, discuss accidents that occurred when they were not followed.

3.4 CHANGES MADE BECAUSE THE REASON FOR EQUIPMENT OR PROCEDURES HAS BEEN FORGOTTEN

This is one of the most common causes of accidents and I have discussed them at length in my book *Lessons from Disaster — How Organisations Have No Memory and Accidents Recur* [24] and suggested actions that could improve corporate memories (see also Section 16.10).

A recycle stream was found to contain a contaminant which produced a runaway reaction if its concentration was high enough. The stream was therefore routinely analyzed. Several years later, after a change in management, the analysis was stopped. A few months later, an explosion occurred [25].

Responsibility was shared, I suggest, between the original supervisor or supervisors who never documented the reasons for the tests in a readily accessible form (if they documented them at all) and the new supervisor or supervisors who stopped the test without knowing why it had been started. **NEVER STOP OR CHANGE A PROCEDURE UNLESS YOU KNOW WHY IT WAS INTRODUCED. NEVER STOP USING EQUIPMENT UNLESS YOU KNOW WHY IT WAS PROVIDED.**

In medieval England there were officials called Remembrancers whose job was to remind the king's courts of matters that they might otherwise forget [26]. (The job still exists but the duties are now ceremonial.) Every process plant needs such a person.

Sections 8.1, 8.2, 8.5, 8.7, 8.9, 14.7, and 14.8 describe other accidents that would not have occurred if the results of changes had been foreseen.

REFERENCES

1. Lees, F.P. (1996). *Loss Prevention in the Process Industries*, 2nd ed., vol. 2, Butterworth-Heinemann, Oxford, UK and Woburn, MA, Chapter 21.

2. Kletz, T.A. (1998). *What Went Wrong? — Case Histories of Process Plant Disasters*, 4th ed., Gulf, Houston, TX, Chapter 2.

3. Sanders, R.E. (1998). *Process Safety — Learning from Case Histories*, Butterworth-Heinemann, Oxford, UK and Woburn, MA.

4. Palmer, P.J. (2003). Private communication.

5. Anon. (1998). *Chemical Manufacturing Incident*, Investigation Report No. 1998-06-NJ, US Chemical Safety and Hazard Investigation Board, Washington, D.C.

6. Fore, H. (1990). Contributions of chemistry to food consumption in *Milestones in 150 Years of the Chemical Industry*, Royal Society of Chemistry, London.

7. Lovelock, J. (2000). *Homage to Gaia — The Life of an Independent Scientist*, Oxford University Press, Oxford, UK, p. 41.

8. Lovelock, J. (2000). *Homage to Gaia — The Life of an Independent Scientist*, Oxford University Press, Oxford, UK, p. 176.

9. Anon. (2001). *Management of Change*, Safety Bulletin No. 2001-04-SB, Chemical Safety and Hazard Investigation Board, Washington, D.C.

10. Lawson-Hall, G. (2001). A costly cost-saving modification. *Loss Prevention Bulletin*, Aug., 160:10–11.

11. Anon. (2002). Explosion in an odorant plant. *Loss Prevention Bulletin*, Dec., 168:19–20.

12. Anon. (1996). Failure to manage can lead to costly losses. *IRI Sentinel*, 1st Quarter, pp. 3–4.

13. Anon. (1983). Thermal insulation on valve causes expansion of flange bolts. *Petroleum Review*, Oct., p. 34.

14. Anon. (2001). Creeping change. *Loss Prevention Bulletin*, Oct., 161:8–9.

15. Kletz, T.A. (2000). *By Accident — A Life Preventing them in Industry*, PFV Publications, London, p. 33.

16. Pickup, R.D. (2001). Dust explosion case study: "Bad things can still happen to good companies." *Process Safety Progress*, **20**(3):169–172.

17. Schofield, M. (1988). Corrosion horror stories. *The Chemical Engineer*, March 446:39.

18. Anon. (1985). Modification malady. *Chemical Safety Summary*, **56**(221):6.

19. Fishwick, T. (2002). Distortion of the end of a full drum of chlorine. *Loss Prevention Bulletin*, Dec., 168:15–16.

20. Anon. (1888). *Railway Passenger Travel 1825–1880*, Scribners, New York; Reprinted by Americana Review, New York.

21. Milner, R. (1991). The yellow fever panic of 1905. *Natural History*, July, p. 52.

22. Ford, R. (2002). Gauge corner cracking — privatisation indicted. *Modern Railways*, **59**(640):19–20.

23. Pugsley, A.G. (1966). *The Safety of Structures*, Arnold, London (quoted by N.R.S. Tait, (1987). *Endeavour*, **11**(4):192).

24. Kletz, T.A. (1993). *Lessons from Disaster — How Organisations have No Memory and Accidents Recur*, Institution of Chemical Engineers, Rugby, UK and Gulf, Houston, TX.

25. Knowlton, R.E. (1990). Dealing with the process safety "Management gap." *Plant/Operations Progress*, **9**(2):108–113.

26. Anon. (2001). *Encyclopaedia Britannica*, Entry on Remembrancer.

Chapter 4

Changes in Organization

The best of systems is no substitute for experience, or for seeing and listening and sniffing for yourself

> — Financial News, *Daily Telegraph* (London), Feb. 16, 2002

The explosion at Flixborough, UK in 1974 drew the attention of the oil and chemical industries to the need to control changes to plants and processes. Many publications (see Chapter 3) have both described accidents that occurred because no one foresaw the results of such changes and suggested ways of preventing such accidents in the future. Only in recent years, however, have we realized that changes in organization can also have unforeseen effects and should likewise be scrutinized systematically before they are made. In some countries, this is now a legal requirement in high hazard industries.

A common organizational change is to eliminate a job and distribute the jobholder's tasks among other workers. Although the jobholder is asked to list all his or her duties, sometimes one or two are missed, especially those carried out by custom and practice and not listed in any job description. For example, someone may have built up a reputation as a "gatekeeper," — someone who knows how to get things done — for example, where scarce spare parts may be squirreled away and so on. Another person may be the only mechanic who really understands the peculiarities of a certain machine. Only after they have left are their distinctive contributions really recognized. Such changes and potential consequences of them are easily missed.

The following are some examples of the unforeseen effects of changes in organization. Sections 3.3.2 and 15.5 describe another.

4.1 AN INCIDENT AT AN ETHYLENE PLANT

The plant was starting up after a turnaround. At 2 AM on the day of the incident, the shift team started the flow of cold liquid to the demethanizer column. A level should have appeared in the base of the column 2 h later. It did not, but problems elsewhere distracted the shift team and they did not notice this until 7 AM. By this time the temperature at the top of the column was −82 °C (−115 °F) instead of the usual −20 °C (−4 °F) and the level in the reflux drum rose from zero to full scale in 10 min. This should have told the shift team that the column had flooded, overflowed into the reflux drum, and would now be filling the flare knock-out drum (see Figure 4.1). However, neither of the two high-level indicator/alarms on this drum, set at 8% and 22% of capacity, showed any response.

It was 12 noon before anyone had a thorough look at the plant. They then found that the wires leading to the column level indicator were dis-

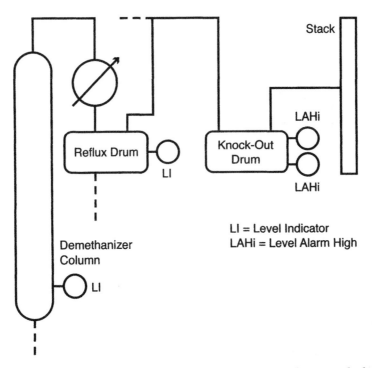

Figure 4-1. The level indicator on the column and the level alarms on the knock-out drum were out of order. The column filled with liquid, which overflowed into the drums and then into the stack.

connected and that the valves between the knock-out drum and its level indicators were closed. (Section 7.1 describes a similar incident.) Both vessels were shrouded with scaffolding and the state of the wires and valves was not easily seen. Liquid was now entering the flare stack. It failed as the result of low-temperature embrittlement but, fortunately, the leaking liquid did not catch fire. No one was injured.

The immediate causes of the incident were the failures to recommission the level instruments before start-up and the slowness of the shift teams to realize what was happening. The underlying causes were far deeper and were due to both short-term and long-term changes in organization.

4.1.1 Short-Term Changes

It was the practice at the plant to work 12-h shifts instead of the usual 8-h shifts during start-ups so that there were more people present than during normal operation. On this occasion the operators refused to do so (though they were willing to work overtime if necessary; this would give them more pay than working 12-h shifts). However, the foremen and shift managers worked 12-h shifts. They changed shift at 7 AM and 7 PM while the operators changed at 6 AM, 2 PM, and 10 PM. This pattern of work destroyed the cohesion that had been built up over the years within each shift and lowered the competence of the team as a whole. A report in the local newspaper said that, "A major influence over the behavior of the operating teams was their tiredness and frustration." A trade union leader was quoted as saying that the management team members were more tired than the operators as they were working 12-h shifts.

In addition to the usual shift personnel, two professional engineers were also present on each shift but their duties were unclear. Were they there to advise the shift manager or, being more senior in rank, could they give him instructions? Should they try to stand back and take an overview or should they get involved in hands-on operations? On the day of the incident, they did the latter and got involved in the details of the problems which distracted everyone from the demethanizer problem.

4.1.2 Long-Term Changes

So far I have followed the published report on the incident [1] but there had also been more fundamental changes. The incident shook the company. It had a high reputation for safety and efficiency and the ethylene plant was considered one of its flagships — one of the least likely places where such a display of incompetence could occur, so what went wrong?

About 7 yr earlier, there had been a major recession in the industry. As in many other chemical companies, drastic reductions were made in the number of employees, at all levels, and many experienced people left the company or retired early. This had several interconnected results.

- Operating divisions were merged and senior people from other parts of the company, with little experience of the technology, became responsible for the ultimate control of some production units.

- There was pressure to complete the turnaround and get back on line within three weeks. This pressure came partly from above but also from within the production and maintenance teams, as the members were keen to show what they could do. They should have aborted the shutdown to deal with the problems that had distracted everyone during the night but were reluctant to do so.

- There were fewer old hands who knew the importance, when there were problems, of having a look around and not just relying on the information available in the control room. A look around would have shown ice on the demethanizer column.

- Delayering had produced a large gap in seniority between the manager responsible for the ethylene plant and the person above him. This made it more difficult for the ethylene manager to resist the pressure to get back on line as soon as possible. Previously, an intermediate manager had acted as a buffer between the operating team and other departments, and he prevented commercial people and more senior managers from speaking directly to the start-up team. In addition, he would probably have aborted the start-up. Senior officers, not foot soldiers, order a retreat.

The company had an outstanding reputation for openness but was reticent about this incident and no report appeared in the open literature — other than the local newspaper — until about 12 yr later, after the company had sold the plant.

4.2 THE LONGFORD EXPLOSION

On September 25, 1998, a heat exchanger in the Esso gas plant in Longford, Victoria, Australia fractured, releasing hydrocarbon vapors and liquids. Explosions and a fire followed, killing two employees and injuring eight. Supplies of natural gas were interrupted throughout the state of Victoria and were not fully restored until October 14. There was no alternative supply of gas, and many industrial and domestic users were without fuel for all or part of the time the plant was shut down. The accident is described in a detailed official report [2], in a book by Andrew Hopkins that concentrates on the underlying causes [3], and more briefly elsewhere [4, 5].

The purpose of the unit in which the explosion occurred was to remove ethane, propane, butane, and higher hydrocarbons from natural gas by absorbing them in "lean oil." The oil, now containing light hydrocarbons and some methane and known as "rich oil," was then distilled to release these hydrocarbons and the oil, now lean again, was recycled. The heat exchanger that ruptured was the reboiler for the fractionation column. The cold rich oil was in the tubes and was heated by warm lean oil in the shell.

As a result of a plant upset, the lean oil pump stopped. There was now no flow of warm lean oil through the heat exchanger and its temperature fell to that of the rich oil, −48 °C (−54 °F). The official report describes in great detail the circumstances that led to the pump stopping. However, all pumps are liable to stop from time to time and the precise reason why this pump stopped on this occasion is of secondary importance. Next time it will likely stop for a different reason. In this case, one of the reasons was the complexity of the plant. It had been designed to recover as much heat as possible and this resulted in complex interactions, difficult to foresee, between different sections.

Ice formed on the outside of the heat exchanger when the flow of warm oil stopped but no one realized that the low temperature was hazardous. Despite long service, the operators had no idea that the heat exchanger could not withstand low temperatures and thermal shocks and that restarting the flow of warm lean oil could cause brittle failure. More seriously, some of the supervisors and even the site manager, who was away at the time, did not know this. It was not made clear in the operating instructions.

The operators' ignorance does not surprise me. When I worked in production, before I became involved full-time in safety, I learned that some operators' understanding of the process was limited. Trouble-shooting depended on the chargehands (later called assistant foremen) and foremen, assisted by those operators who were capable of becoming chargehands or foremen in the future. In recent years, I have heard many speakers at conferences describe the demanning and empowerment their companies have carried out and wondered whether the operators of today are really better than those I knew in my youth. At Longford they were not.

Esso claimed that their operators had been properly trained and that there was no excuse for their errors. But the training emphasized that knowledge the operators needed to do their job rather than an understanding needed to deal with unforeseen problems. They were tested after training but only for knowledge, not for understanding. One operator was asked why a certain valve had to be closed when a temperature fell below a certain value. He replied that it was to prevent thermal damage and received a tick for the correct answer. At the inquiry [3], he was asked what he meant by thermal damage and replied that he "had no concept of what that meant." When pressed, he said that it was "some form of pipework deformity" or "ice hitting something and damaging pipework." He had no idea that cold metal becomes brittle and could fracture if suddenly warmed.

Now we come to crucial changes in organization. Two major changes were made during the early 1990s. In the first, all the engineers, except for the plant manager, the senior man on site, were moved to Melbourne. The engineers were responsible for design and optimization projects, and for monitoring rather than operations. They did, of course, visit Longford from time to time and were available when required but someone had to recognize the need to involve them.

In the second change, the operators assumed greater responsibility for plant operations and the supervisors (the equivalent of foremen) became fewer in number and less involved. Their duties were now largely administrative.

Both of these changes were part of a company-wide initiative by Exxon, the owners of Esso Australia and were the fashion of the time. There was much talk of empowerment and reduced manning. The report concluded that, "The change in supervisor responsibilities . . . may have contributed by leaving operators without properly structured supervision." It added, "Monthly visits to Longford by senior management failed to detect these shortcomings and were therefore no substitute for essential on-site supervision."

On the withdrawal of the engineers, the report said that it

"appears to have had a lasting impact on operational practices at the Longford plant. The physical isolation of engineers from the plant deprived operations personnel of engineering expertise and knowledge, which previously they gained through interaction and involvement with engineers on site. Moreover, the engineers themselves no longer gained an intimate knowledge of plant activities. The ability to telephone engineers if necessary, or to speak with them during site visits, did not provide the same opportunities for informal exchanges between the two groups, which are often the means of transfer of vital information."

None of this was recognized beforehand. Chats in the control room and elsewhere allow operators to admit ignorance and discuss problems in an informal way that is not possible when a formal approach has to be made to engineers at the company headquarters. Empowerment can become a euphemism for withdrawal of support.

Anecdotal of course, but supportive of the real-life work problems caused by too much formality: on one occasion when I was a safety adviser with ICI Petrochemicals Division, I was asked to move my small department to a converted house just across the road from the division office block. I objected as I felt that this would make contact with my colleagues a little bit harder because they would be less likely to drop by our offices.

At Longford there were also errors in design. The heat exchanger that failed could have been made from a grade of steel that could withstand low temperatures or a trip could have isolated the flow of cold liquid if the temperatures of the heat exchanged fell too far. These features were less common when the plant was built than they became 30 years later but they could have been added to the plant. The designs of old plants should be reviewed from time to time. This is particularly important if they have undergone changes that were not individually studied. We cannot bring all old plants up to all modern standards — inconsistency is the price of progress — but we should review old designs and decide how far to go. Esso intended to Hazop the plant but the study was repeatedly postponed and ultimately forgotten. Another design weakness was the overly complex heat recovery system already mentioned.

Exxon has a high reputation for its commitment to safety and for the ability of its staff. Was Longford a small plant in a distant country that fell below the company's usual standards or did it indicate a fall in standards in the company as a whole? Perhaps a bit of both. Exxon did not require Esso Australia to follow Exxon standards and the Longford plant fell far below them. Exxon was fully aware of the hazards of brittle failure (see Section 6.3) but their audit of Esso did not discover the ignorance of this hazard at Longford. On the other hand, the removal of the engineers to Melbourne and the reductions in manning and supervision were company-wide changes. It also seems that in the company as a whole the outstandingly low lost-time accident rate was taken as evidence that safety was under control. Unfortunately, the lost-time accident rate is not a measure of process safety.

Esso was prosecuted for 11 failures "to provide and maintain so far as is practicable for employees a working environment that is safe and without risk to health" and had to pay the largest fine ever imposed by the State of Victoria for such an offense. However, the fine was small compared with the claims for damages caused by the loss of natural gas. In a summary and review [6] of the trial, Hopkins says that it produced no new causal insights. However, it provided an object lesson in how not to handle the defense in such a case. Hopkins concludes that "Esso's decision to plead not guilty, its conduct at the trial and its refusal to accept responsibility led the Judge to conclude that the company had shown no remorse, and the absence of corporate remorse weighed heavily in his decision not to mitigate the penalties in any way."

Although Esso claimed at the inquiry that an operator was responsible for the accident, they did not claim this at the trial, perhaps because the attempt to blame the operator had produced adverse publicity. It is unusual today for managers to blame the person whose triggering action is the last in a long series of missed opportunities to prevent an accident (see Chapter 16). Perhaps the decision to blame the operator was made by a lawyer who knew nothing about plant operation or human nature.

4.3 OUTSOURCING

A marketing manager in a company that manufactured ethylene oxide foresaw a market for a derivative. The company operated mainly large continuous plants while the production of the derivative required a batch plant. The derivative was needed quickly and the company did not want to spend capital on a speculative venture. The manager therefore looked for a contract manufacturer who could make it for them. He found one able to undertake the task and signed a contract without consulting any of his technical colleagues. The manufacturer was quite capable but unfortunately was located in a built-up area. When it was realized that ethylene oxide was being handled there, this gave rise to some concern even though the stock on site was quite small.

A few years later the buildings in a considerable area around the plant were demolished as part of a slum clearance project. The regulators then refused permission for new ones to be built in their place. Before they could develop the site, the local authority had to pay the contract company to move its plant to a new location.

This incident occurred some years ago, before present-day regulations came into force. It probably could not happen today, but it is a warning that outsourcing of products or services is a change that should be systematically considered before it takes place.

4.4 MULTISKILLING AND DOWNSIZING

Multiskilling presents specific problems, illustrated by the Flixborough explosion. The site was without a mechanical engineer for several months, as the only one — the works engineer — had left and his successor had not arrived. Arrangement had been made for a senior engineer from one of the owner companies to be available when needed but the unqualified engineers who designed and built the temporary pipe that failed did not realize that these tasks were beyond their competence and did not see the need to consult him [7]. Similarly, in many plants there is now no longer an electrical engineer but the control engineer is responsible for electrical matters. An electrical engineer is available for consultation somewhere in the organization but will the control engineer know when to consult him? *Will he know what he doesn't know*?

The same applies at lower levels. Will the process operator who now carries out simple craft jobs be able to spot faults that would be obvious to a trained craftsman?

One of the underlying causes of the collapse of a mine tip at Aberfan in South Wales, which killed 144 people, most of them children, was similar. Responsibility for the siting, management, and inspection of tips was given to mechanical rather than civil engineers. The mechanical engineers were unaware that tips on sloping ground above streams can slide and have often done so [7].

On downsizing, according to one report [8],

No one likes to talk about it, but having less experienced people working in increasingly sophisticated computer-generated manufacturing operations increases the risks of serious and costly mistakes. The investigation into an explosion in one US chemical plant [in 2001] found that the engineer in charge has only been out of college a year, and the operators in the control room at the time of the accident all had less than a year's experience in the unit. Not surprisingly, the explosion was attributed to operator error. . . . And even when errors are not caused by inexperience, diagnosing and fixing them often takes longer when veteran employees are no longer around to help.

Attributing the errors to the operators is, of course, superficial. The underlying cause is either downsizing or employment conditions that failed to retain employees.

4.5 ADMINISTRATIVE CONVENIENCE VERSUS GOOD SCIENCE

My final example, from James Lovelock's autobiography, *Homage to Gaia* [9], shows what happened when "administrative convenience ruled and good science and common sense came second," though the results were a decline in effectiveness rather than safety. He was working in a government-funded research center that employed chemists and biologists. It was amalgamated with another similar institution some distance away. To the administrators it seemed sensible to move all chemists to one site and all biologists to the other, as this would avoid the need to duplicate the services each group required. The administrators did not realize, and did not listen to those who did, the numerous research benefits gained from informal day-to-day contact between people from different disciplines. Both institutions declined. As we saw in Section 4.2 on the Longford incident, a similar loss of communication occurred when the professional engineers were moved from a plant to the company's head office.

4.6 THE CONTROL OF MANAGERIAL MODIFICATIONS

As with changes to plants and processes, changes to organization should be subjected to control by a system, which covers the following points:

- Approval by competent people. Changes to plants and processes are normally authorized by professionally qualified staff. The level at which management changes are authorized should also be defined.
- A guide sheet or check list. Hazard and operability studies are widely used for examining proposed modifications to plants and processes before they are carried out. For minor modifications, several simpler systems are available [10]. Few similar systems have been described for the examination of modifications to organization [11]. Some questions that might be asked by those who have to authorize them are suggested here:
  - Each modification should be followed up to see if it has achieved the desired end and that there are no unforeseen problems or failures to maintain standards. Look out for near misses and for failures of operators to respond before trips operate. Many people do not realize that the reliability of trips is fixed on the assumption that most deviations will be spotted by operators before trips operate. We would need more reliable trips if this were not the case.
  - Employees at all levels must be convinced that the system is necessary or it will be ignored or carried out in a perfunctory manner. A good way of doing this is to describe or, better, discuss, incidents such as those described in the foregoing, and which occurred because there was no systematic examination of changes.

4.6.1 Some Points a Guide Sheet Should Cover

Define what is meant by a change: Exclude minor reallocations of tasks between people but do not exclude outsourcing, major reorganizations following mergers or downsizing, or high-level changes such as the transfer of responsibility for safety from the operations or engineering director to the human resources director. Accidents may be triggered by people but are best prevented by better engineering [12].

Nearly half of the companies that replied to a questionnaire on the management of change said that they included organizational change under this rubric [13]. However, they may not include the full range of such changes.

Some questions that should be asked include:

- How will we assess the effectiveness of the change over both the short- and the long-term?
- What will happen if the proposed change does not have the expected effect?
- Will informal contacts be affected (as at Longford)?
- What extra training will be needed and how will its effectiveness be assessed?
- Following the change, will the number, knowledge, and experience of people be sufficient to handle abnormal situations? Consider past incidents in this way.
- If multiskilling is involved, will people who undertake additional tasks know when experts should be consulted? See Section 4.4 on multiskilling.

Except for minor changes, these questions should be discussed by a group, as in a hazard and operability study, rather than answered by an individual. *None* or *not a problem* should not be accepted as an answer unless backed up by the reasons for the answer. Any proposal for control of changes in organization should be checked against a number of incidents, such as those described herein, to see if it could have prevented them.

REFERENCES

1. Anon. (1999). A major incident during start-up. *Loss Prevention Bulletin*, Dec., 156:3.

2. Dawson, D.M., and J.B. Brooks (1999). *The Esso Longford Gas Plant Explosion*, State of Victoria, Australia.

3. Hopkins, A. (2000). *Lessons from Longford*, CCH Australia, Sydney, Australia.

4. Kletz, T.A. (2001). *Learning from Accidents*, 3rd ed., Butterworth-Heinemann, Oxford, UK and Woburn, MA, Chapter 24.

5. Boult, D.M., R.M. Pitblado, and G.D. Kenney (2001). Lessons learned from the explosion and fire at the Esso gas processing plant at

Longford, Australia, *Proceedings of the AIChE 35th Annual Loss Prevention Symposium*, April. 23–25.

6. Hopkins, A. (2002). Lessons from Longford — The Trial, *Journal of Occupational Health and Safety — Australia and New Zealand*, Dec., **118**(6):1–72.

7. Kletz, T.A. (2001). *Learning from Accidents*, 3rd ed., Butterworth-Heinemann, Oxford and Woburn, MA, Chapters 8 and 13.

8. De Long, D.W. (2002). *Research Note: Uncovering the Hidden Costs of Lost Knowledge in Global Chemical Companies*, Accenture Institute for Strategic Change, Cambridge, MS, p. 2.

9. Loveock, J. (2000). *Homage to Gaia*, Oxford University Press, Oxford, UK, p. 301.

10. Lees, F.P. (1996). *Loss Prevention in the Process Industries*, 2nd ed., vol. 2, Butterworth-Heinemann, Oxford, UK and Woburn, MA, Chapter 21.

11. Conlin, H. (2002). Assessing the safety of process operation staffing arrangements. *Hazards XVI — Analysing the Past, Planning the Future*, Symposium Series No. 148, Institution of Chemical Engineers, Rugby, UK, pp. 421–437.

12. Philley, J. (2002). Potential impacts to process safety management from mergers, downsizing, and re-engineering. *Process Safety Progress*, **21**(2):151–160.

13. Keren, N., H.H. West, and M.S. Mannan (2002). Benchmarking MOC practices in the process industries. *Process Safety Progress*, **21**(2):103–112.

This chapter is based in part on a paper presented at the Hazards XVII Conference held in Manchester, UK in March 2003, and is included with the permission of the Institute of Chemical Engineers.

Chapter 5

Changing Procedures Instead of Designs

Rigid, repetitive behavior, resistance to change and a lack of imagination are common symptoms.

— Extract from an article on autism, *Daily Telegraph* (London), April 29, 2002

When we join an organization, and especially when we are young, we tend to follow, and are expected to follow, its ways of thinking and acting. It is usually only later, when we have gained experience, that we start to question these default actions. This chapter describes a common, but unfortunate, way many organizations react after an accident.

There are several different actions we can take after we have identified a hazard (as a result of an accident or in some other way) to prevent it from causing another accident or to mitigate the consequences if it does: Our first choice, whenever "reasonably practicable," should be to remove the hazard by inherently safer design. For example, can we use a safer material instead of a toxic or flammable one? Even if we cannot change the existing plant, we should note the change for possible use on the next plant. (*Reasonably practicable* is a UK legal phrase that recognizes the impracticability of removing every hazard and implies that the size of a risk should be compared with the cost of removing or reducing it, in money, time, and trouble. When there is a gross disproportion between them, it is not necessary to remove or reduce the risk [1].)

If we cannot remove the hazard, then our second choice should be to keep it under control by adding passive protective equipment, that is, equipment that does not have to be switched on or does not contain moving parts. The third choice is active protective equipment, that is, equipment switched on automatically; unfortunately, the equipment may be neglected and fail to work or it may be disarmed.

The fourth choice is reliance on actions by people, such as switching on protective equipment; unfortunately, the person concerned may fail to act, for a number of reasons, such as forgetfulness, ignorance, distraction, poor instructions, or, after an accident, because he or she has been injured. Changes to procedures instead of designs are often called *work arounds*.

Finally, we can use the techniques of behavioral science to improve the extent to which people follow procedures and accepted good practice. By listing this as the last resort, I do not intend to diminish its value. Safety by design should always be our aim but it is often impossible, and experience shows that behavioral methods can create substantial improvement in the everyday types of accident that make up most of the lost-time and minor accident rates. However, the technique has little effect on process safety. Behavioral methods should not be used as an alternative to the improvement of plant design or methods of working when these are reasonably practicable.

To clarify various ways of preventing incidents, let us consider a simple but common cause of injury and even death, particularly in the home — falls on the stairs.

The inherently safer solution is to avoid the use of stairs by building a single-story building or using ramps instead of stairs.

If that is not reasonably practicable, a passive solution is to install intermediate landings so that people cannot fall very far or to avoid types of stairs, such as spiral staircases, which make falls more likely. An active solution is to install an elevator. Like most active solutions, it is expensive and involves complex equipment that is liable to fail, expensive to maintain, and easy to neglect.

The procedural solution is to instruct people always to use the handrails, never to run on the stairs, to keep them free from junk, etc. This can be backed up by behavioral techniques: specially trained fellow workers (or parents in the home) look out for people who behave unsafely and tactfully draw their attention to the action.

Similarly, if someone has fallen into a hole in the road, as well as asking why it was not fenced or someone removed the fence or if the lighting should be improved, we should ask if there is a reasonably practicable alternative to digging holes in the road. Could we drill a route for pipes or cables under the road or install culverts for future use when roads are laid out? Must we run pipes and cables under the road instead of overground?

In some companies, the default action after an accident is to start at the wrong end of the list of alternatives and recommend a change in procedures or better observation of procedures, often without asking why the procedures were not followed. Were they, for example, too complex or unclear or have supervisors and managers turned a blind eye in the past? Changing procedures is, of course, usually quicker, cheaper, and easier than changing the design, but less effective. This chapter describes some accidents in which changes in design would have been less expensive but nevertheless only changes in procedures were made. The first two accidents could easily have killed someone; the third is trivial, but they all illustrate the same point. There are other examples in Sections 1.4 and 13.2.

Designers today often consider inherently safer options but the authors of incident reports do so less often. The very simplicity of the idea seems to make it hard for some people to grasp it. Perhaps they are expecting something more complex or — and this is perhaps more likely — it goes against the widely accepted belief that accidents are someone's fault and the job of the investigation is to find out whose. Having identified the culprit, we are less likely to blame him or her than in the past; we realize that he or she may not have been adequately trained or instructed, and that everyone makes occasional slips, but nevertheless his or her action or inaction caused the incident. In some companies, they blame a piece of equipment. It is hard for some people to accept that the incident is the result of a widespread and generally accepted practice in design and operations.

5.1 MISLEADING VALVE LAYOUTS

5.1.1

The fine adjustment valve A in Figure 5-1 had to be changed. The operator closed the valve below it. To complete the isolation, he intended to close the valve on the other side of the room in the pipe leading to valve A. He overlooked the double bends overhead and closed valve B, the one opposite valve A. Both of the valves that were closed were the third from the ends of their rows. Note that the bends in the overhead pipes are in

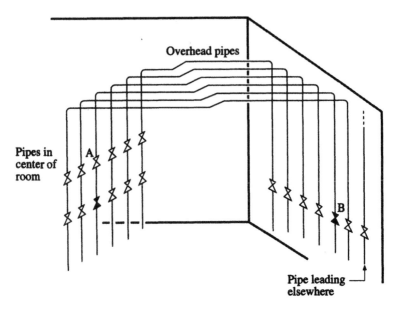

Figure 5-1. To change valve A, the operator closed valves B and C. From reference 5. Reprinted with the permission of the Institution of Chemical Engineers.

the horizontal plane. When the topwork of valve A was unbolted, the pressure of the gas in the line caused the topwork to fly off and hit the wall, fortunately missing the mechanic who had unbolted it.

The report on the incident recommended various changes in the instructions to make the duties of people who prepare equipment for maintenance clearer than they were. They were told to trace lines to make sure that the correct isolations have been made.

Color coding of the pipes or valves would have been much more effective but was not considered. The default action of many of the people in the company was to look first for changes in procedures, to consider changes in design only when changes in procedure were not possible, and to consider ways of removing the hazard rather than controlling it only as the last resort.

The ideal solution, of course, would be to rearrange the pipework so that valves in the same line were opposite each other. To do so in the existing plant would be impracticable but the point should be noted for the future. After a similar incident elsewhere, a design engineer once said to me that it was difficult enough to get all the pipework into the space available without having to worry about such fine points as the relative positions of valves. This may be so but putting valves in unexpected positions leads to errors.

The changes made after the accident were not even the most effective procedural ones. The incident could have been given widespread publicity, not just immediately afterward but regularly in the future, and made part of the training of people authorized to issue permits-to-work.

5.1.2

Figure 5-2 shows a similar situation. To save cost, three waste heat boilers shared a common steam drum. Each boiler had to be taken off line from time to time for cleaning. On two occasions, the wrong valve was closed (D3 instead of D2) and an on-line boiler was starved of water and

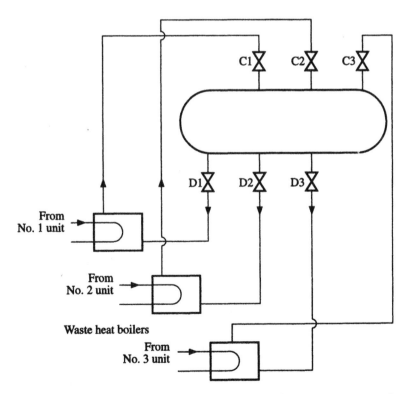

Figure 5-2. Note positions of isolation valves on the common steam drum. Reprinted with permission of *EVHE*, Fig. 2.6, 2001, page 23.

overheated. The chance of an error was increased by the lack of labeling and the arrangement of the valves — D3 was below C2. On the first occasion the damage was serious. High temperature alarms were then installed on the boilers. On the second occasion they prevented serious damage but some tubes still had to be changed. A series of interlocks were then installed so that a unit had to be shut down before a key could be removed; this key was needed to isolate the corresponding valves on the steam drum.

A better design, used on later plants, is to have a separate steam drum for each waste heat boiler (or group of boilers if several can be taken off line together). There is then no need for valves between the boiler and the steam drum. This is more expensive but simpler and free from opportunities for error. Note that we do not begrudge spending money on complexity but are reluctant to spend it on simplicity.

It is obviously impracticable to change the layout of the existing valves but perhaps color coding would have been sufficient to prevent further errors. It would have been simpler and cheaper than the mechanical interlocks.

5.2 SIMPLE REDESIGN OVERLOOKED

A bundle of electric cables was supported by cable hangers. The hooks on the ends of the cable hangers were hooked over the top of a metal strip (Figure 5-3 top). The electric cables had to be lowered to the ground to provide access to whatever lay behind them and then replaced. They were put back as shown in the second part of Figure 5-3. This increased the load on the upper hooks. One failed, thereby increasing the load on the adjacent ones and then they also failed. Altogether, a 60 m (200 ft) length of cables fell down [2].

Many people would fail to see this hazard. Training is impracticable if, as is probably the case, many years will pass before the job has to be done again. The best solution is to use cable hangers strong enough to carry the weight even if they are used incorrectly.

5.3 UNIMAGINATIVE THINKING

Wash-basins filled with water were installed at a plant so that anyone splashed with a corrosive chemical could wash it off immediately. The basins were covered to keep the water clean but people used the covers

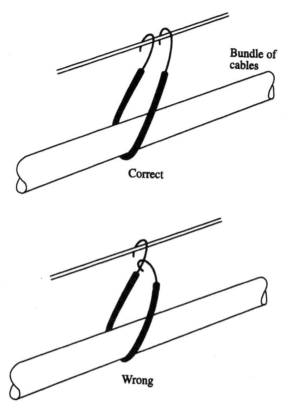

Figure 5-3. Different ways of supporting a bundle of cables. Reprinted with permission of EVHE, Fig. 10.2, 2001, p 177.

as tables (Figure 5-4a). Figure 5-4b shows the action taken and Figure 5-4c shows a better solution.

Perhaps there is something wrong with our educational system and/or company culture when educated and professionally trained people take the action shown in Figure 5-4b.

5.4 JUST TELLING PEOPLE TO FOLLOW THE RULES

A tank containing high-level radioactive liquid was fitted with instruments for measuring density and level. They were purged with steam at intervals. Before opening the steam valve, the operator was instructed to check that there was steam in the line by measuring the temperature of a

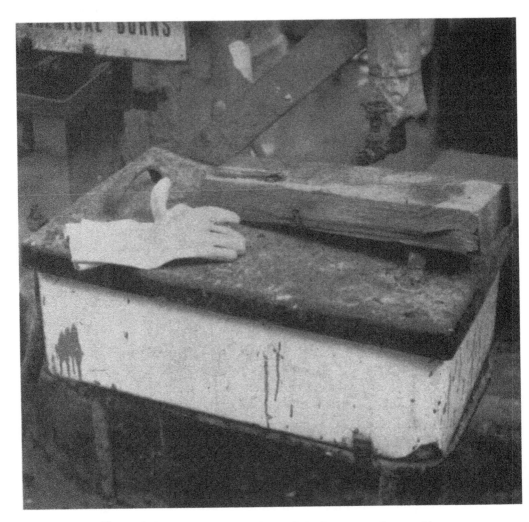

Figure 5-4a. The cover over the wash basin was used as a table.

steam trap and checking that it was over 93 °C (200 °F). However, he merely felt the trap and finding it was hot, he opened the steam valve. Unknown to him, the steam line had been isolated 16 h beforehand. (Presumably conduction from beyond the isolation valve kept the trap hot.) As the steam cooled, it developed a vacuum and this sucked the radioactive liquid into the steam line. Radioactive alarms sounded and fortunately no one received a significant dose.

Figure 5-4b. The solution.

The report [3] drew attention to failures to follow procedures: The people who drained and isolated the steam line did not inform those responsible for purging the instruments; and the operator who was asked to carry out the purging was not adequately trained because he had never done the job before but only watched other people do it.

The report recommended that managers should stress the proper use of procedures, that before carrying out a task operators should stop, think about the task, the expected response, and the actions required if it failed to occur, and so on. There was no suggestion that the procedures could be improved, for example, by fitting a warning notice on lines that are out-of-use, or that the design could be improved. It is surprising that there was no check valve in the steam line. They are not 100% reliable but can greatly reduce the size of any back flow. Check valves with moving parts would be difficult to maintain in a radioactive environment but fluidic ones would be suitable. Another possibility is a catchpot to catch any liquid that does flow into the steam line.

5.5 DON'T ASSEMBLE IT INCORRECTLY

When an accident occurs because construction or maintenance workers assemble equipment incorrectly, the default action of many managers is

Figure 5-4c. A better solution.

to tell them to take more care in future and to check that it has been assembled correctly. Or perhaps they provide training on the correct method of assembly. They do not realize that equipment should be designed so that it cannot be assembled the wrong way. Even when it is impractical to change the design of existing equipment, we should at least ask the design organization to use a better design in the future.

During rough weather, water entered the engine room of a fishing vessel through the intake of the ventilation fans. It fell onto the switchboard and a short circuit set it afire; the fire was soon extinguished but all power was lost. The crew was unable manually to fully close the doors through which the nets were pulled onboard and water entered through these doors. The ship had to ask for help and was towed back to port.

Why did water enter through the ventilation intake? The louvers in it had been installed incorrectly so that they directed spray and rain into the engine room rather than away from it. (The report said that they had been installed upside down but the authors must have meant back-to-front.) See Figure 5-5.

The report's first recommendation [4] was that louvers should be checked to make sure they are fitted correctly. It did not suggest that they should be designed so that they could not be fitted incorrectly, so that it was obvious if they were or so that the inside and outside, top and bottom,

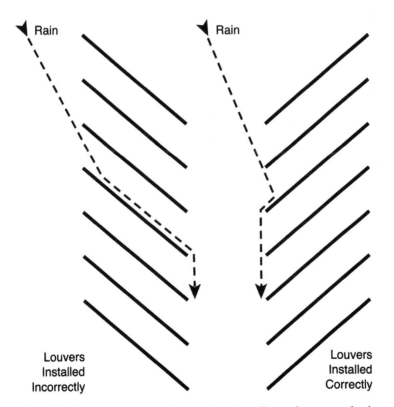

Figure 5-5. The louvers were installed so that they directed spray and rainwater through them.

were clearly labeled. However, the report did recommend that switchboards should be covered to prevent water entering from above.

5.6 TIGHTEN CORRECTLY OR REMOVE THE NEED

A hose was fastened to its connector with a type of clip used for radiator hoses in cars (known as Jubilee clips in the UK). The connection leaked. The recommendation in the report on the incident was, "Check tightness of Jubilee clips during maintenance." These clips are not robust enough for industrial use and a better recommendation would have been to replace them with bolted clips.

Similarly, a steel plate fell from a clamp while being lifted because the bolt holding it in position was not tightened sufficiently. The incident was classified as a human failing and the operator was told to be more careful in future. It would have been better to use a type of clamp that is not dependent for correct operation on someone tightening it to the full extent [5].

5.7 SHOULD IMPROVEMENTS TO PROCEDURES EVER BE THE FIRST CHOICE?

Improving procedures is often the only possible choice, but are there times when it is more effective than changing designs? This may be the case with road accidents. Up to the late 1970s, the United States had the lowest fatal accident rate per thousand vehicles in the world. The figure has continued to fall and is now about half the rate it was then, a considerable achievement. But other countries have done even better and the US is now 13th in the road safety league, behind the United Kingdom, Sweden, the Netherlands, Germany, Canada, Australia, Japan, and several other countries, but not France. The better performance of these countries is not due to better vehicle design, as they all use similar vehicles and the US tends to use heavier and, therefore, safer cars. Nor is there a significant difference in the design of roads. Leonard Evans [6] suggests that significant differences in the countries with lower accident rates are a more restrictive alcohol policy, which is enforced more rigorously, stricter enforcement of seat belt laws, and prohibition of the sale and use of radar detectors. If so, further improvement in the US depends on better enforcement of procedures.

If you work in the process industries, the most dangerous task your employer asks you to perform may be to drive between sites.

REFERENCES

1. Health and Safety Executive. (2001). *Reducing Risks, Protecting People*, HSE Books, Sudbury, UK.

2. Anon. (1991). Worker injured by falling power and data cables. *Operating Experience Weekly Summary*, No. 99-08, Office of Nuclear and Facility Safety, US Dept of Energy, Washington, D.C., p. 1.

3. Anon. (1999). Radioactive tank contents contaminate steam line. *Operating Experience Weekly Summary*, No. 99-34, Office of Nuclear and Facility Safety, US Dept of Energy, Washington, D.C., pp. 11–13.

4. Anon. (2001). Upside down louvres. *Safety Digest — Lessons from Marine Accident Reports*, No. 2/2001, Marine Accident Investigation Branch of the UK Department of Transport, Local Government and the Regions, London, p. 40.

5. Kletz, T.A. (2001). *An Engineer's View of Human Error*, 4th ed., Institution of Chemical Engineers, Rugby, UK, p. 43.

6. Evans, L. (2002). Traffic crashes. *American Scientist*, **90**(3):244–253.

Chapter 6

Materials of Construction (Including Insulation)

Lay not up for yourself treasure upon earth: where the rust and moth doth corrupt.

— Book of Common Prayer

At school we all knew people who were often present when there was any trouble but they were rarely identified as the major culprit. In preparing the index for the 4th edition of my book, *What Went Wrong — Case Histories of Process Plant Disasters*, I was surprised to find that certain words appeared, often as secondary or incidental causes, much more often than I expected. I expected to find (and did find) frequent references to fires, explosion, pumps, tanks, modifications, and maintenance, but was surprised how many references there were to rust, insulation, and brittle failure. These incidents and some others are described in what follows. There are more details in *WWW* (and elsewhere when another reference is quoted).

6.1 RUST

6.1.1 Rust Formation Uses up Oxygen

A tank was boxed up with some water inside. Rust formation used up oxygen and three men who entered the tank were overcome; one of them died. No tests were carried out before the men were allowed to enter the tank as it had contained *only water*. In another similar incident, three men were sent to inspect the ballast tanks on a barge at an isolated wharf. The first man to enter collapsed and a second man who tried to rescue him also collapsed; one of them died. In a third incident, rust formation caused a tank to be sucked in. Rusting is usually slow but can be rapid under certain conditions; it increases rapidly when the humidity is high. (Chapter 2 describes other accidents in confined spaces.)

6.1.2 Rust-jacking

To avoid welding in a plant that handled flammable liquids, an extension was bolted onto a pipebridge. The old and new parts were painted but water penetrated the gap between the bolted surfaces and rusting occurred. Rust has seven times the volume of the iron from which it is formed and it forced the two surfaces apart. Some bolts failed and a steam main fractured. Fortunately, the pipes carrying flammable gases did not fail [1].

Similarly, corrosion of reinforcing bars in concrete can cause the concrete to crack and break away.

6.1.3 Liquid Can be Trapped Behind Rust

The roof on an old gasoline tank had to be repaired by welding. The tank was emptied and steamed, and tests showed that no flammable vapor was present. However, the tank had been made by welding overlapping plates together along the outside edge only, a method no longer used. Some gasoline was trapped by rust in the space between the overlapping plates. Welding vaporized it and ignited the vapor; it blew out the molten weld and singed the welder's hair.

Another similar accident had worse results. Heavy oil trapped between overlapping plates was vaporized and exploded. The roof of the tank was lifted. One man was killed and another badly burned.

6.1.4 Rust as Catalyst

Rust can initiate the polymerization of ethylene oxide at ambient temperature. Once the temperature reaches 100 °C (212 °F), the reaction becomes self-sustaining and may lead to explosive decomposition. An explosion in an ethylene oxide distillation column may have been started by rust, which had accumulated in a dead-end space. Rust on the inside surface of a tank can promote the polymerization of other substances. Even if the liquid has been treated with an inhibitor, this will not prevent polymerization of vapor, which can condense on the walls or roof [2].

A vigorous reaction between chlorine and a steel vaporizer, described as burning, led to a leak and the loss of a ton of chlorine. The steam was supplied to the vaporizer at a gauge pressure of ≈ 1 bar (15 psi) and a temperature of $\approx 100\,°C$ (212 °F), so it was not nearly hot enough to ignite the chlorine–iron reaction, which starts at ≈ 200–250 °C (390–480 °F). The bottom of the vaporizer was found to be packed with scale containing 80% iron oxide. It is possible that this material had catalyzed the reaction. However, it is more likely that the culprit was traces of methanol, which had been used to clean the vaporizer and had not been thoroughly purged afterwards. Methanol, like many other organic compounds, can react with chlorine and generate enough heat to start the chlorine–iron reaction [3].

6.1.5 Rust Jams a Valve

A chlorine cylinder was left standing, connected to a regulator, for 8 months. The valve rusted and seemed to be fully closed, though it was not. When someone disconnected the regulator, gas spurted into his face. Four people were hospitalized.

6.1.6 Thermite Reactions

If rusty steel is covered by aluminum paint (or smeared with aluminum in any other way) and then hit by a hard object, such as a hammer, a thermite reaction can occur: the iron oxide reacts with the aluminum to form aluminum oxide and iron. A temperature of 3,000 °C (5,400 °F) can be reached and this can ignite any flammable gas, vapor, or dust that is present.

A thermite reaction can also occur between rust and any other metal that has a greater affinity for oxygen. A fractionation column was packed with bundles of 0.1-mm-thick corrugated titanium sheets that had become coated with a layer of rust only $25\,\mu$ thick. During a shutdown, it was decided to check that the correct construction materials had been used. This was done by passing a grinding wheel lightly and quickly across the surface of components and noting the characteristics of the sparks. They ignited the titanium and set off a thermite reaction. The fire spread rapidly, causing extensive damage [4] (see also Section 10.1).

6.1.7 Rust Formation Weakens Metal

Some handheld fire extinguishers are fitted with rubber or plastic feet to protect the bases. If water enters the space between the foot and the extinguisher, it can cause rusting. In at least two cases, extinguishers have ruptured while in use and killed the person who was holding them. Manufacturers advise users to remove the plastic feet and check for corrosion yearly but it would be better not to use extinguishers of this type [5].

The 7-in diameter exit pipe from the superheater of a steam boiler was threaded and screwed into a flange. The gauge pressure was 17 bar (250 psi). The joint leaked causing substantial damage to both pipework and the building roof. The investigation showed that seepage of steam along the threads of the screwed joint had caused corrosion and that all the gaps in the threads had contained rust. Many of the grooves were full of it. There was evidence that the two threads were never tightly engaged and that there were gaps between them from the start. The report [6] recommended that screwed joints should not be used on large pipes, say, those over 2 in diameter. However, many companies do not allow them at all except for low-pressure cold water lines and for small bore instrument lines after the first isolation valve and then only for nonhazardous materials. The report also suggested that existing screwed joints might be opened for inspection every few years. (See also Section 12.5.)

6.1.8 Old Plants and Modern Standards

The boiler described in the last incident was a very old one and raises the question, how far should we go in bringing old plants up to modern standards. Some changes are easy, for example, installing gas detectors for the detection of leaks. Some are impossible, such as increasing the spacing between different parts of the unit. In between there are changes that are possible but expensive, such as replacing pipework by grades of steel that are less likely to corrode or can withstand lower temperatures. Some companies carry out a *fitness for purpose* study of such suspect equipment, replacing some, radiographing or stress-relieving some, fitting extra measurements, alarms, or trips on some, or training operators to pay particular attention to the operating conditions [7]. If a *fitness for purpose* study had been carried out on the boiler after the incident, the conclusion would, I think, have been to replace the screwed joints by flanged ones.

6.1.9 Stainless Steel Can Rust

Stainless steel can rust if it is exposed to particularly aggressive conditions, either physically (for example, by cleaning with steel wool or wire brushes) or chemically [8].

In all these cases, rust was not the major culprit. Proper procedures should be followed before vessels are entered, equipment should be designed without pockets in which liquids or rust can collect, coated or stainless steel should be used if rust can affect materials in contact with it, flammable mixtures should not be tolerated except under rigidly defined conditions where the risk of ignition is accepted, cylinder valves should not be left open for months, and people should be made aware of these hazards and of the properties of rust. (See also Sections 1.6.2 and 12.1.)

6.2 INSULATION

Insulation, like rust, is often mentioned in incident reports, though rarely as the major culprit. It has many benefits but we should be aware of its drawbacks.

6.2.1 Insulation Hides What is Beneath It

On several occasions, small diameter branches have been covered by insulation, overlooked, and not isolated before maintenance. Short tags on blinds (slip-plates) may not be noticed on insulated lines. In one incident, a check (nonreturn) valve was hidden by insulation and a new branch was installed on the wrong side of it. As a result, a relief valve was bypassed and equipment was overpressured and it ruptured.

Most important of all, insulation can hide corrosion. The commonest cause of corrosion beneath insulation is ingress of water, especially water contaminated with acids or with chlorides, which can cause stress-corrosion cracking of stainless steel. Sections of insulation should be removed periodically for inspection of the metal below, making sure that no gaps are left when it is restored. During inspection, special attention should be paid to places where corrosion is likely, such as insulation supports and stiffening rings which can trap water, gaps in the insulation around nozzles, and insulation around flanges and valves. Nonabsorptive insulation should be used when possible. Make sure insulation is not left lying around where it can get wet before installation. Remember that while warm equipment may dry out wet insulation, the rate of corrosion doubles for every 15–20 degrees C (27–36 degrees F) rise in temperature. Reference [9] reviews the subject (See also Section 12.2).

Corrosion and a leak of propylene took place beneath insulation on equipment that had been in use for 15 yr. The corrosion had occurred only on those parts of the unit that operated between about 0 and 5 °C (32–40 °F). These parts were frequently wetted by condensation from the atmosphere. Some of the equipment was replaced with stainless steel and the rest was inspected more frequently [10].

Supports can corrode as well as equipment. The corroded legs of a 2,000 m³ (530,000 gal) LPG sphere collapsed during a hydrotest, when it was 80% full of water. One man was killed and another seriously injured. It was then found that:

- Water had penetrated the gap between the concrete insulation and the legs as the cap over the concrete was inadequate.
- There were also vertical cracks in the concrete.
- Repairs to the concrete had not adhered to the old concrete, leaving further gaps
- The deluge system had been tested with sea water.
- Inspection was inadequate.

Underlying all these problems, according to the report, was a poor maintenance system, poor management, and ignorance of what could occur and what precautions should be taken.

The sand foundation below a fuel oil tank subsided. This was not noticed as the insulation came right down to the ground. Water collected in the space that was left and caused corrosion. The floor of the tank collapsed and 30,000 tons of hot oil came out. The bottom 0.2 m (8 in) of the tank walls should have been left free of insulation so that they could be inspected easily.

6.2.2 Wet Insulation is Inefficient

If insulation is allowed to get wet, it not only encourages corrosion but also loses much of its efficiency: 4% moisture by volume can reduce the thermal efficiency by 70% as water has a thermal conductivity of up to 20× greater than most insulation materials [11].

6.2.3 Spillages on Insulation Can Degrade and Ignite

When organic liquids are spilled on hot insulation, they can degrade and their auto-ignition temperatures can fall by 100–200 degrees C (180–360 degrees F). In one incident, ethylene oxide leaked through a hairline crack in a weld on a fractionation column onto insulation and reacted with moisture to form polyethylene glycols. When the metal covering on the insulation was removed in order to gain access to an instrument, air leaked in and the polyethylene glycols ignited. The fire heated a pipe containing ethylene oxide. It decomposed explosively and the explosion traveled into the fractionation column, which was destroyed. A leak of ethylene oxide from a flange may have caused a similar incident on another column.

On another unit, a spillage onto insulation was the result of filling the heat transfer section with oil until it overflowed. A month later the oil caught fire; the flames caused a leak of gas, which exploded, causing further damage. Solution: contaminated insulation should be removed promptly.

6.2.4 Some Insulation is Flammable

Large tanks are sometimes insulated with plastic foam, which is lighter and cheaper than non-flammable insulating materials. However, the foams can, and do, catch fire.

6.2.5 Metal Coatings over Insulation Should be Grounded

When a glass distillation column cracked, water was sprayed onto it to disperse the leak of flammable vapor. The water droplets were charged and the charge collected on the metal insulation cover, which was not grounded. A spark was seen to jump from the insulation cover to the water line, but fortunately it did not ignite the leak.

6.2.6 Insulation Can Fall Off

If 10% of thermal insulation falls off (or is removed for maintenance or inspection and not replaced), then we lose 10% of its effect. However, if 10% of fire insulation is missing, we lose all the effect as the bare metal will overheat and fail. Missing insulation should be replaced promptly.

6.3 BRITTLE FAILURE

This is a third subject often mentioned in accident reports. A famous case is discussed in Section 4.2.

6.3.1 Temperature Too Low as a Result of Adiabatic Cooling

Most materials become brittle if they are cooled below the brittle-ductile transition temperature. This is the commonest cause of brittle failure and often occurs when the pressure on a liquefied gas is reduced. Vessels, heat exchangers, and road tankers containing liquefied petroleum gas have been cooled below their transition temperatures by deliberate or accidental venting and have then failed when subjected to a sudden shock. We should use construction materials that can withstand foreseeable reductions in temperature outside normal operating conditions.

6.3.2 Temperature Too Low as a Result of Adding Cold Fluids

A vessel broke into 20 pieces when it was filled with cold nitrogen gas from a liquid nitrogen vaporizer. Vehicle tires have exploded in contact with liquid nitrogen. A large pressure vessel failed during a pressure test at the manufacturer's as the water used was too cold. A piece weighing 2 tons went through the workshop wall and traveled 15 m (50 ft).

6.3.3 Manufacturing Flaws

As a result of a flaw during manufacture 40 yr earlier, a tank containing 15,000 tons of diesel oil opened up like a zipper. For most of those 40 yr, the tank had been used to store warm fuel oil and the high temperature prevented brittle failure. The flaw could have been spotted if the tank had been adequately radiographed. A liquid carbon dioxide vessel failed catastrophically as result of poor quality welding. The triggerring event was the failure of a heater that was intended to prevent evaporative cooling.

6.3.4 Use of Unsuitable Materials

Cast iron is brittle and cannot withstand sudden shocks. A 6-in cast iron steam valve failed spectacularly when subjected to water hammer.

A 2,000 m³ (530,000 gal) propane tank opened up like a zipper as it was not made from a crack-arresting material. The designers had assumed incorrectly that cracks could be prevented and, unfortunately, when one occurred it spread rapidly. The cause of the crack may have been attack of a weld by bacteria in the seawater used for pressure testing followed by a poor repair. It is difficult to be certain that a crack will *never* occur. It is good practice to prevent the spreading of any that do occur by using grades of steel that can withstand the temperatures reached during normal and abnormal operation.

6.4 WRONG MATERIALS OF CONSTRUCTION

6.4.1 Wrong Materials of Construction and Contaminants

316L stainless steel was specified for a fractionation column and its connecting pipework. Seven years later, the bottom section was replaced with a taller one so that the column could be used for a different purpose. A few months later, leaks occurred in some of the old connecting pipework. It was then found that it had been made from 304L steel instead of 316L. This did not matter in its previous use, but an acidic byproduct formed and attacked the 304L steel as a result of the newer use.

During the 1970s, many incidents occurred because the wrong grade of steel was supplied (see *WWW*, Section 16.1). Many companies introduced materials identification programs: every piece of steel entering the site — pipes, flanges, and welding rod as well as complete items of equipment — was tested to check that it was made of the material specified. Many of these programs were abandoned when suppliers were able to show that the quality of their procedures had met approved standards. I question if this is wise as many suppliers do not seem to understand that supplying a "similar" grade of steel instead of the one specified can have serious results (see also Section 9.7).

In addition, in the 304L/316L stainless steel incident just described, the new shell also leaked and some tray supports gave way. This was traced to chlorine in the steam used for cleaning the column between runs. It is well known that chlorine can cause stress corrosion cracking of stainless steel but we do not expect to find chlorine in steam. It had picked up the chlorine from deposits in the base of the column and carried them into the column. There were, unknown to everyone, traces of chlorine ions in the feedstock [12].

Sometimes, of course, the wrong construction material is specified. When some vessels made from 316L stainless steel corroded in contact with strong acids, they were replaced in 317L, a grade that is usually more resistant. However, at the same time the temperature was raised from 90 °C to 110 °C (195–230 °F). Corrosion still occurred. The plant chemist noticed that the amount of iron, nickel, and chromium in the product had increased but this was not recognized as evidence of corrosion [12].

Unforeseen corrosion in heat exchangers in a sugar refinery was traced to a combination of a less-than-ideal material of construction, presence of chlorine in an additive, and copper carried over from an earlier part of the process. Metallic contamination in the food industry is a well-recognized problem and must be closely monitored. The corrosion took place when the exchangers were being cleaned with acids to remove deposits, and although the plant had been hazard-and-operability studied, it was not clear that the cleaning process had been included in the study. It is, in effect, a different process carried out using the same equipment and should have been the subject of a separate study [13].

6.4.2 A Hasty Reaction When the Plant Leaked

Nitric acid leaked through a plug on a ring main, which normally operated at a gauge pressure of 7 bar (100 psi). It caused some corrosion of the equipment on which it dripped. The plug was one of several fitted by the design and build contractor in case it was found necessary to install additional instruments, though this is not certain as no one who worked on the plant was involved during design. It seems that the company was never consulted about the need for screwed plugs. None of them had been unscrewed during the 8 yr that had elapsed since the plant was built. The plug that leaked was made from mild steel though the pipework was stainless steel and was incorrectly seated. A polytetrafluoroethylere (PTFE) wrapping around the threads prevented an earlier leak.

It is sometimes necessary to install temporary plugs to aid pressure testing, to assist draining or, as in this case, to make it easier to install additional instruments. Their positions should be registered and if it becomes clear that they are not needed, they should be welded up. However, do not seal weld over an ordinary screwed plug as if the thread corrodes the full pressure inside the equipment is applied to the seal. Use a specially designed plug with a full strength weld. Also, do not use sealing compounds in joints that are going to be welded as the welding will vaporize the sealing compound and make the weld porous. If sealing compound has been used, the joint should be cleaned before welding. (For other incidents see Section 12.5 and *WWW*, Sections 7.1–7.5.)

When the leak was discovered, the process supervisor immediately decided to drain the ring main via several different drain points so that it was emptied as soon as possible. He did not open the vent at the highest point in the ring main. The opening of several drain points produced complex pressure and vacuum transients in the ring main and unpredictable movements of slugs of liquid. As a result, a column of liquid 3 m (10 ft) high was discharged from a vent point. Afterwards the plug was never found so it may have been sucked into the ring main by a transient vacuum. In total, 100 liters (25 US gal) of nitric acid were spilled.

The supervisor's action was quite understandable but once the pressure in the ring main was reduced, a short delay would not have mattered. Twenty minutes spent discussing possible methods would have been time well spent.

There may also have been some air pockets in the ring main and a revised filling procedure was adopted when the ring main was refilled. Nevertheless, another pressure discharge occurred from an open vent during refilling. This shows how difficult it is to estimate the pressures developed when complex pipework — there were many changes of elevation — is being filled or emptied. It is equally difficult to avoid such features during design. There was no standing instruction on how to drain the ring main. Complex systems have complex problems and their causes are much more difficult to understand than outsiders realize.

The investigation disclosed that although nitric acid was used infrequently, nevertheless the ring main was kept up to pressure at all times. It need not have been. Was a ring main really needed?

According to the company report, the most important lesson was not to rush into action but I think a more important one is the need to plan ahead for jobs that will have to be done sooner or later and not leave the people on the job to improvise when the time comes.

Another lesson is that we should question the need for every plug and look out for plugs that design or construction staff insert for their own convenience. Those added by construction staff are usually not shown on any drawing.

6.5 CORROSION SENDS A COLUMN INTO ORBIT

Corrosion occurred in an absorber tower, 18 m (60 ft) tall and 2.6 m (8.5 ft) diameter, in which liquid monoethanolamine removed hydrogen sulfide from gaseous propane and butane. After four years' service, the base of the column was replaced without any post-welding heat treatment. Two years later a Monel liner was fitted to reduce corrosion but it did not cover the repair weld. After another 8 yr, a circumferential crack formed. In places it extended nine-tenths of the way through the 1-in thick wall. Once it broke through, it grew rapidly and the upper part of the column landed over a kilometer away (see Figure 6-1). The escaping gas was ignited, perhaps by a welder's torch, and exploded. Gasoline tanks were damaged and the contents ignited; the flames impinged on a liquefied petroleum gas tank which ruptured, producing a boiling liquid expanding vapor explosion (BLEVE). Seventeen people were killed and damage was extensive.

The investigation showed that the welding of the new bottom section, without any post-welding heat treatment, had produced a hard microstructure which was susceptible to hydrogen attack and brittle failure [14,15,16].

Figure 6-1. The result of an absorber failure. From reference 18. Reprinted with the permission of Gulf Professional Publishing.

Unfortunately, as so often happens, the published reports give no indication of the underlying reasons for the managerial failings. Did the company have any material scientists on their staff? Did they hire an inexperienced contractor and leave it to him? Did the senior managers believe that every welder is capable of welding everything? The incident is a warning to companies who think that knowledge and experience are inessential luxuries, that it is okay to be a naïve client and leave everything to a contractor. Elsewhere [17] I have described many accidents that occurred, from the nineteenth century to the present day, because companies placed too much trust in contractors.

Stress corrosion cracking is common in amine gas absorption columns. Reference [18] recommends polymer coating of construction materials.

6.6 UNEXPECTED CORROSION

Corrosion of a pipe led to a leak of >2 tons of a mixture of gaseous chlorine, hydrochloric acid, and hydrogen fluoride. As soon as the leak was detected, the affected section of the plant was isolated from the rest by remotely operated emergency isolation valves but there was no valve of any sort between the leaking pipe and a vessel. The leak was stopped after two hours when a fitter, wearing full protective clothing and an air mask, and standing on a ladder, succeeded in clamping a rubber sheet over the leak.

The plant was seven years old. The process materials were known to be corrosive, the most suitable materials of construction were used and a life of five years was expected. The vessels were inspected regularly and some had been replaced but the pipe that failed had never been inspected or renewed. It seems odd to inspect vessels regularly but never inspect the pipes connected to them. What do you do?

The acid gases caused considerable damage to the electronic control equipment. The cost of replacing them and the affected pipework was too great and the plant was demolished.

The protective clothing used during the emergency was rarely used and much of it was found to be in poor condition and unusable. All emergency equipment should be scheduled for regular inspection.

6.7 ANOTHER FAILURE TO INSPECT PIPEWORK

Many companies that inspect pipes carrying hazardous materials do not inspect those that carry nonhazardous ones, but that does not mean that

Figure 6-2. This low-pressure steam main was not scheduled for regular inspection and the damage was undetected.

they never fail. Figure 6-2 shows an anchor on a low-pressure steam main with axial expansion joints (bellows) on both sides. The damage was due to operation at a higher temperature than design and probably occurred months or even years before it was noticed. All expansion joints should be registered for regular inspection.

6.8 HOW NOT TO WRITE AN ACCIDENT REPORT

An operator noticed a small leak of hot nitric acid vapor from a pipe. It seemed to be coming from a small hole on a weld. Radiography showed that there was significant corrosion of the weld and of a condenser just below it. A temporary patch was fitted to the leak and plans were made to replace the condenser and pipe at a turnaround scheduled to take place a few months later.

Full marks to the company for writing a report on the incident and circulating it widely within the company — but the report left many questions unanswered:

- How long had the equipment been in use?
- Had it been radiographed previously?
- Was the original welding to the standard specified?
- Was a positive materials identification program in force when the plant was built and were the pipe, welding rods, and condenser checked to make sure that they were made from the correct grade of steel? Was a suitable grade specified?

REFERENCES

1. Schofield, M. (1988). Corrosion horror stories. *The Chemical Engineer*, March, 446:39–40.

2. Lees, F.P. (1996). *Loss Prevention in the Process Industries*, 2nd ed., Butterworth-Heinemann, Oxford, UK and Woburn, MA, pp. 11/91 and 22/58.

3. Fishwick, T. (1998). A chlorine leak from a vaporiser. *Loss Prevention Bulletin*, Feb., 139:12–14.

4. Anon. (2002). A fire in titanium stucture packing. *Loss Prevention Bulletin*, Aug., 166:6–8.

5. Anon. (2000). *Safety Alert on a Fire Extinguishing Fatality,* Mary Kay O'Connor Process Safety Center, College Station, TX, 2 Oct.

6. Fishwick, T. (1999). Failure of a superheater outlet joint resulting in equipment damage. *Loss Prevention Bulletin*, April, 146:5–7.

7. Kletz, T.A. (2001). *Learning from Accidents*, 3rd ed., Butterworth-Heinemann, Oxford, UK and Woburn, MA, Section 30.7.

8. Elliott, P. (1998). Overcome the challenge of corrosion. *Chemical Engineering Progress*, **94**(5):33–42.

9. Posteraro, K. (1999). Thwart corrosion under industrial insulation. *Chemical Engineering Progress*, **95**(10):43–47.

10. Lindley, J. (2002). In brief — corrosion. *Loss Prevention Bulletin*, Feb. 163:5–6.

11. Collier, K. (2002). Insulation. *Chemical Engineering Progress*, **98**(10):47.

12. Lyon, D. (2002). Suspected weld failures in process equipment. *Loss Prevention Bulletin*, Feb. 163:7–9.

13. Fabiano, B. (2002). Corrosion in a sugar refinery. *Loss Prevention Bulletin*, Feb. 163:10–12.

14. Anon. (2002). Union Oil amine absorber fire. *Loss Prevention Bulletin*, Feb. 163:20–21.

15. Ornberg, R. (1984). On the Job — Illinois, *Firehouse*, Oct. pp. 74–76, 176.

16. Vervalin, C. H. (1987). Explosion at Union Oil. *Hydrocarbon Processing*, **66**(1):83–84.

17. Kletz, T.A. (2001). *Learning from Accidents*, 3rd ed., Butterworth-Heinemann, Oxford, UK and Woburn, MA, Chapters 5, 13.7, 16.5, and 22.

18. Mogul, M.G. (1999). Reduce corrosion in amine gas absorption columns. *Hydrocarbon Processing*, **78**(10):47–56.

Chapter 7

Operating Methods

Human nature will instinctively modify what should be done into what can be done especially if this makes the job easier or keeps the job moving in some way.

> — Anon., *Loss Prevention Bulletin*, Oct. 2000

7.1 THE ALARM MUST BE FALSE

We all know of occasions when operators have said, "The alarm must be false" and sent for the instrument technician. For example, the high-level alarm on a storage tank operates. The operator knows the tank is empty and ignores the alarm. By the time the technician arrives, the tank is overflowing. Someone has left a valve open and the liquid intended for another tank has flowed into the first one.

Here is an incident where the operator had a good reason for thinking that the alarm was false. The three reactors on a plant were being brought back on line after a turnaround. Number 1 had been stabilized to normal operating conditions but Nos. 2 and 3 were still at the early stages of start-up. The temperature on No. 2 started to rise and the high-temperature alarm sounded. It seemed impossible that any reaction could have occurred at so early a stage and all other readings were normal so the operator decided that the instrument must be faulty and sent for a technician. A little while afterwards, a pipe on No.1 reactor was found to be growing red hot. During the shutdown, work had been done on the temperature-measuring instruments on the three reactors and the leads from Nos. 1 and 2 were accidentally interchanged. (Section 4.1 describes a similar error.)

It is a good practice to test all trips, interlocks, and alarms after a shutdown, or at least those that have been maintained. The incident also shows the value of a walk around the plant when anything is out of the ordinary. We may not know what we are looking for, but we never know what we may find.

Someone had a rather similar experience after collecting her car following the repair of some minor accident damage. On the way home she had to make several turns and on each occasion other cars hooted her. When she got home, she found that the rear direction indicators had been connected up the wrong way round so that when she signaled a left turn, the right indicators flashed. When she telephoned the repair company, they at first insisted that they always checked direction indicators to make sure that they were wired correctly. In fact, only the front ones had been checked.

7.2 A FAMILIAR ACCIDENT —
BUT NOT AS SIMPLE AS IT SEEMED

Moving liquid into the wrong vessel is one of the commonest accidents in the chemical industry and is usually, and often unfairly, blamed on an error by the operator. An unusually frank and detailed report shows the superficiality of such a view.

Some liquid had to be transferred from one vessel to another. Such movements were common though transfers between the two vessels involved on this occasion were unusual. The foreman asked an experienced operator to carry out the operation but before this operator could do so he was called to a problem elsewhere on the unit and the job was left to a new and inexperienced operator, with another experienced operator keeping an eye on him from time to time.

The trainee went to the transfer pumps where there was a diagram of the pipework. At one time all the vessels and valves were numbered. Unfortunately, painters had painted over many of the labels, which had then been removed as illegible and never replaced. The trainee opened a wrong valve. As a result, the liquid went into a vessel that was out of use and ready for refurbishment. Some of the liquid leaked out of a faulty thermocouple pocket. About 50 liters (13 gal) of a corrosive liquid were spilled inside a building and some of it dripped down to the floors below.

The trainee checked that the level in the suction vessel was falling but he could not check that the liquid was arriving in the intended delivery vessel as other streams were entering at the same time. In addition, the level indicator and alarm on the vessel into which the liquid was actually being pumped had been disconnected as the vessel was out of use.

7.2.1 What Can We Learn?

- If a job had to be left to an inexperienced man, was manning adequate or had downsizing gone too far? A few years later, following a more serious incident, manning was increased.

- In a piece of unfortunately common *management-speak*, the report blamed the operators for not reporting the missing labels. Of course, they should have reported them, but the supervisors and managers (and the auditors) should also have seen and reported them. If the operators had reported the missing labels, would they have been replaced? This is the sort of fiddly job that maintenance teams often never get around to doing. If operators report this sort of fault and nothing is done, they do not report such faults again. Following the later incident just mentioned, many missing labels were found on other units.

- The pipeline leading to the vessel that was out of use should have been isolated by a blind or at least by a locked valve. The valve handle had been removed but on this plant that merely indicated that the valve was used infrequently, not that it should be kept shut.

- If equipment is not positively isolated, by blinding or disconnection, then its level instrumentation should be kept in operation. The levels in tanks that were supposed to be out of use have often changed.

- If toxic, flammable, or corrosive liquids are liable to leak inside buildings, the floors should be liquid-tight.

The recommendations were followed up on the unit where they occurred, but because the spillage was small, it had little impact elsewhere in the plant and company. This is a common failing. After the tires on a company vehicle were inflated to such a high pressure that they burst, the recommended inflation pressures were painted above the wheels of all the site vehicles, but only in the factory where the burst occurred, not anywhere else.

7.3 MORE RELUCTANCE TO BELIEVE THE ALARM

Another incident made worse because measurements were not available occurred on a ferry boat but has lessons for the process industries. The exhaust gas from the engines was used to raise steam in two waste heat boilers. One of them developed a steam leak and was shut down. The steam lines were isolated and the boiler drained but the exhaust gas continued to pass through the boiler as there was no way of bypassing it. The high-temperature alarm on the boiler sounded but nothing wrong could be found. The inspection — and another carried out when the vessel reached port and the engines had been stopped — could not have been very thorough as when the engines were restarted a few hours later, an expansion joint (bellows) was found to be glowing red hot. The passengers were told to leave the ship and the fire service was called.

The lack of a bypass was a weakness in the design but the ship's crew seems to have been too ready to believe that the alarm was false (as described in Section 7.1). The inspection carried out in the port may not have been thorough because shutting down the engines disabled the alarms and the crew may not have known this (compare Section 7.2.1, penultimate bullet).

Once the hot expansion joint was found, the incident was dealt with correctly and efficiently. In my experience, the same is true in the process industries. Poor design and operation may have led to an incident but once it occurs, the right action is usually taken. In my time at the plant, when the fire alarm sounded, maintenance workers left the area — rightly as they were not trained to deal with fires — while operators ran towards it. Most fires were extinguished even before the site fire service arrived.

The report on this incident comes from a periodic review of marine accident reports [1]. Most of them are of nautical incidents such as ships running aground or colliding with other ships but a surprising number are process incidents, such as the one just described, unsafe entries to tanks, a foam-over (See Section 8.12) because hot oil was put into a tank containing a water layer, a fire in a galley because butane from an old aerosol can leaked into a cupboard, choked vents, and many failures of lifting gear. Some other marine incidents are described in Section 15.2.

7.4 THE LIMITATIONS OF INSTRUCTIONS

However many instructions we write, we never think of everything and so people should be given the knowledge and understanding they need to handle situations not covered in the instructions. This is usually illustrated by descriptions of complex accidents such as the nuclear accident at Three Mile Island [2] in Middletown, PA, US, but a very simple incident illustrates the same theme.

The plant handled a very toxic material. When filter cartridges contaminated with this material were changed, the old ones were placed in sealed plastic bags and taken to another building for cleaning and disposal. If the bags were dropped, they might easily rupture and so instructions stated that the bags must be moved on a trolley. The trolleys were conveyed downstairs in the elevator.

What would you do if you were asked to move a bag and the elevator was out of order (or you were his foreman)? The person who wrote the instructions never foresaw this problem.

The man asked to move the bag did as many people would have done: he carried the bag downstairs. He could then have put it on a trolley but having carried it so far he carried it the rest of the way to the foreman's office and put it on the table. The bag slid off and punctured and the room had to be evacuated and cleaned.

The inquiry brought to light the fact that the operators and the foremen did not fully appreciate the hazards of the material on the filters. People will follow instructions to the letter only when they understand the reasons for doing so. We do not live in a *Charge of the Light Brigade* society in which people will unthinkingly obey every command (see also Sections 2.5.2, 8.12, and 14.5).

The inquiry also revealed that bags containing contaminated filters had been carried downstairs on at least two other occasions when the elevator was out of order, but nothing was said. Perhaps the foreman preferred not to know or more likely, he never brought together in his mind the two contradictory facts: the elevator was out of use and a bag had gotten downstairs. (Section 14.6 describes an incident in which a computer "believed" two contradictory facts.)

7.5 THE LIMITATIONS OF INSTRUCTIONS AGAIN

Thirty gallons of sludge were being pumped into a 55-gal drum. To avoid splashing, two operators fitted the lid on the drum. They did not realize that with no vent for the air to escape, the pressure in the drum would rise. After a while, they noticed that the flow had stopped and that the drum was bulging. They then realized what had happened and decided to remove the lid. As one of them was doing so, the lid flew off, injuring the operator and splashing him with a toxic sludge.

The report [3] emphasized the need to prepare better instructions and hazard check lists for all jobs but, as stated in the previous item, we cannot cover every possibility in our instructions and the longer we make them, the less likely they will be read. We can tell people everything they should do but we cannot tell them everything they should not do. To quote a judgment from the United Kingdom's supreme court, the House of Lords [4], "(A person) is not, of course, bound to anticipate folly in all its forms, but he is not entitled to put out of consideration the teachings of experience as to the form that those follies commonly take." We could replace folly by *human error*.

Accidents such as the one just described are best prevented by better training rather than better instructions, that is, by giving people an understanding of basic scientific principles, in this case that if something is put into a vessel either something, usually air, has to get out or the pressure will rise (but if water is put into a vessel containing a soluble gas, such as ammonia, the pressure will fall).

7.6 EMPTY PLANT THAT IS OUT OF USE

An official report [5] drew attention to a hazard that is easily overlooked. A vessel was not used for several months. It had been pumped dry but unknown to the operators, a layer of solid residue had been left behind in the vessel. When the vessel was brought back into use, on the same duty as before, the fresh reactants reacted with the residue, causing a rise in temperature and the emission of gas into the working area.

The report recommended that:

• Whenever possible, equipment that is going to be left out of use for longer than usual should be emptied completely.

• If that cannot be done (or has not been done), then the material remaining should be tested with the materials that are to be added to see if there is any unforeseen reaction.

• In some cases, it may be possible to prevent deterioration of residues by covering them with a layer of water or other solvent.

7.7 A MINOR JOB FORGOTTEN — UNTIL THERE WAS A LEAK

A solution of a very toxic liquid was kept in a storage tank fitted with a cooling coil, which developed a leak. As cooling was not necessary for the liquid now stored in the tank, a blind was fitted to the cooling water outlet line and it was decided to cap the inlet line. This could not be done immediately because of pressure of other work; therefore, for the time being the inlet line was kept up to pressure so that water leaked into the tank rather than the reverse. The capping job was repeatedly postponed and ultimately forgotten.

Five years later, the water pumps were shut down for a short time and the drop in pressure allowed some of the liquid in the tank to enter the water line. A leak occurred from a sample valve on the water inlet line near the tank — the variation in pressure may have caused the sample valve to leak — and some of the toxic solution leaked out. The leak went into a duct underneath the sample valve. From there it should have flowed to a drain. However, instead it flowed down a *temporary* line, not shown on any drawing, that the construction team had installed and never removed, and dripped down the building. Some water was poured down in an attempt to dilute and sweep away the leak but its effectiveness was doubtful.

7.7.1 What Went Wrong?

It is hardly necessary to say that jobs should not be forgotten. The fact that capping the cooling water outlet line was forgotten suggests the lack of a good safety management system or of the resources necessary to operate and maintain it. This incident was trivial but several years later the company was in trouble with a regulatory authority for its failure to maintain and operate adequate safety management systems. Coming events cast their shadows before they arrive and incidents like this one can serve as warnings that all is not well. Of course, the report said all the right things about the need to log outstanding jobs and so on, and things may have improved at the unit involved, but there was no serious attempt to look at and if necessary change methods elsewhere in the plant and company.

The sample point, and similar ones on other cooling water lines, were rarely used and were removed after the incident. Removing redundant or temporary equipment is one of those jobs that is frequently postponed and then forgotten. Meanwhile the equipment is not maintained and ultimately gives trouble. (See the notes on plugs in Section 6.4.2.) Construction teams should be asked to list any temporary pipes, supports, drains, or plugs that they have installed for their own convenience and have not removed.

7.8 DESIGN ERROR + CONSTRUCTION ERROR + OPERATING ERROR = SPILLAGE

Figure 7-1 shows the layout of the relevant pipework on an experimental unit that was used intermittently. In the original design, the pump seals were supplied with water via valve D. However, the pressure in the main varied and therefore, to provide a more consistent supply, a small head tank was installed and the old supply line to the pump seals was then used as the drain line. During construction, the need to disconnect it from the water main was overlooked and it remained connected. Perhaps the designers of the modification failed to tell the construction team that the line should have been disconnected or perhaps they were told but failed to do so. Either way, nobody checked the job thoroughly (if at all) after completion and the company's procedure for the control of modifications was ignored, as the plant was only an experimental one.

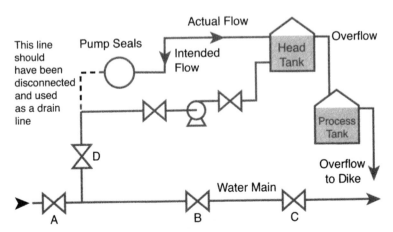

Firgure 7-1. When the unit was modified, the contractors forgot to disconnect the dotted pipeline. As a result, water flowed in the opposite direction to that intended and into the head tank.

Valves A, B, and C were opened to supply water for an hour or so to another part of the unit. Valves B and C were then closed but A was left open. Valve D was already open. Water flowed in the opposite direction to that intended and into the head tank. There was a ballcock valve on the normal inlet but not, of course, on what was intended to be the outlet line so the water filled the head tank and overflowed into the dike. The high-level alarm in the dike sounded but no one heard it. As result of demanning, there was no one present in this part of the plant during the night!

To add to the problem, the head tank did not overflow directly into the dike but into another tank that contained a solution of a process material in water. This tank overflowed so what could have been a spillage of water turned into a spillage of process liquid. The unusual arrangement of the pipework probably arose because the plant was an experimental one so as few pipes as possible were rerouted when the duties of the tanks were changed.

Figure 7-1 looks simple but it shows only a few pipes. There were many more, and many of them followed circuitous routes as the result of earlier modifications and changes of use. The isometric drawing attached to the full report looks like a plate of spaghetti.

7.8.1 What Went Wrong?

- Complexity in pipework (and everything else) leads to errors. Simplicity is worth extra cost. (It is usually cheaper but not always. See Section 5.1.2.)

- Modification control procedures should not be skipped and should be applied to experimental units as well as production plants and to changes in organization, such as demanning, as well as changes to processes (see Chapters 3 and 4).

- Checking of completed pipework should not be left to construction teams, in-house or contracted. The operating team should check thoroughly. They are the ones who suffer the consequences of errors in construction.

REFERENCES

1. Anon. (2002). According to the book! *Safety Digest — Lessons from Marine Accident Reports*, No. 1/2002, Marine Accident Investigation

Branch of the UK Department of Transport, Local Government and the Regions, London, pp. 30–31.

2. Kletz, T.A. (2001). *Learning from Accidents*, 3rd ed., Butterworth-Heinemann, Oxford, UK and Woburn, MA, Chapter 11.

3. Anon. (1999). Propelled drum lid injures research assistant. *Operating Experience Weekly Summary*, No. 99–4, Office of Nuclear and Facility Safety, US Dept of Energy, Washington, D.C., pp. 1–3.

4. Wincup, M. (1966). *The Guardian* (London), 7 Feb.

5. Nuclear Installations Inspectorate (1974). *Report by the Chief Inspector of Nuclear Installations on the Incident in Building B204 at the Windscale Works of British Nuclear Fuels Ltd on 26 September 1973*, Her Majesty's Stationery Office, London.

Chapter 8

Explosions

At 11 minutes past 11 on the morning of November 27th, 1944, the Midlands was shaken by the biggest explosion this country has ever known. 4,000 tons of bombs stored 90 ft down in the old gypsum mines in the area, blew up, blasting open a crater 400 ft deep and $\frac{3}{4}$ mile long. Buildings many miles away were damaged. This pub had to be rebuilt and one farm, with all its buildings, wagons, horses, cattle and 6 people completely disappeared.

> — Notice outside the Cock Inn, Fauld, near Burton-on-Trent, England

The immediate cause of most explosions, that is, violent releases of energy, is an exothermic chemical reaction or decomposition that produces a large amount of gas. However, some explosions such as those described in Sections 8.2, 8.10, and 8.12 have a physical cause. The Fauld explosion is described in Reference [1]. The ignition source was thought to be the rough handling of a sensitive detonator that was being removed from a bomb.

8.1 AN EXPLOSION IN A GAS-OIL TANK

An explosion followed by a fire occurred in a 15,000 m³ (4 million gal) fixed-roof gas oil tank while a sample was being taken; the sampler was killed. The explosion surprised everybody as the gas oil normally had a flash point of 66 °C (150 °F). However, the gas oil had been stripped with hydrogen to remove light materials, instead of the steam originally used, and some of the hydrogen had dissolved in the gas oil and was then released into the vapor space of the storage tank. The change from steam to hydrogen had been made twenty years earlier.

Calculations showed that 90% of the dissolved hydrogen would be released when it was moved to the storage tank and would then only slowly diffuse through the atmospheric vent. Samples were tested for flash point but the hydrogen in the small amounts taken would have evaporated before the tests could be carried out.

When the change from steam to hydrogen stripping was made, it seems that nobody asked if hydrogen might be carried forward into the storage tank. There was no management-of-change procedure in operation at the time.

The source of ignition was probably a discharge of static electricity from the nylon cord used when lowering the sample holder into the tank. Cotton, the recommended material, usually contains enough moisture to conduct electricity while synthetic cords are usually nonconducting. A charge could have built up on the nylon cord as a result of friction between the sampler's glove or cloth and the nylon, while the sample holder was being lowered into the tank, and then discharged to the walls of the tank. No liquid had been moved into the tank during the previous 10 hr so any charge on the gas oil had ample time to discharge.

During the investigation, a 1995 standard for tank sampling was found. It stated, "In order to reduce the potential for static charge, nylon or polyester rope, cords or clothing should be used." A copy of the accident report [2] was sent to the originators of the standard. They replied apologizing for the omission of the word "not"!

A similar accident, another explosion in a tank containing gas oil contaminated with hydrogen, was reported 14 yr earlier [3]. Unfortunately, this incident was not known to anyone in the plant where the second explosion occurred.

8.1.1 Lessons Learned

This explosion, like many others, shows that the only effective way of preventing explosions and fires of gases or vapors is to prevent the formation of flammable mixtures. Sources of ignition are so numerous and the amount of energy needed for ignition is so small (0.02 mJ in this case) that we can never be sure that we have eliminated all sources of ignition. (Energy of 0.02 mJ is the amount released when a one cent coin falls 1 cm. This amount, concentrated into a spark or speck of hot metal, will ignite a mixture of hydrogen and air.)

Nevertheless, we should do what we can to remove sources of ignition. Sample holders lowered into tanks, should be held by conducting cords and in addition, as the holder is lowered, the cord should touch the side of the opening in the tank roof so that any charge generated is removed.

Changes in processes, such as replacing steam by hydrogen, should be Hazop studied as well as changes to plant design. The teams should ask, "What will be the result if any materials present at earlier stages of the process are still present?" "They can't be" is rarely, if ever, an adequate answer.

Note that the crucial change, replacing steam by hydrogen, took place long before the explosion. Twenty years went by before the right combination of circumstances for an explosion arose. This is typical of incidents triggered by static electricity. Note that I did not write *caused*. I regard the change from steam to hydrogen as the cause or, more fundamentally, the lack of adequate study of possible consequences before the change. Like most accidents, this one had happened before.

There is more on static electricity in Sections 2.2.3, 3.2.7, 6.2.5, and 10.7.

8.2 ANOTHER SORT OF EXPLOSION

A tank with a capacity of $726\,m^3$ (200,000 gal) was used for the storage of methylethyl ketone (MEK). The contents were moved into a ship. The transfer pipeline was not emptied immediately afterwards, as it was used frequently for this product, but on the evening of the following day it was decided to empty it by blowing the contents back into the tank with nitrogen at a gauge pressure of 5 bar (75 psi), the usual method.

About 5 min after the level in the tank had stopped rising and before the nitrogen was shut off, there was an explosion in the tank followed by a fire. The roof separated from the walls along half the circumference. As MEK is explosive (flash point $11\,°C/52\,°F$), everyone assumed at first that there had been a chemical explosion. However, there was no obvious source of ignition and static electricity could be ruled out as MEK has a high conductivity. Any static formed will flow to earth through the tank walls in a fraction of a second (as long as the walls are grounded). Someone then asked why the explosion occurred when it did rather than at another time, always a useful question to ask when investigating an accident, especially an explosion. Had anything changed since the last time when the transfer pipeline had been blown with nitrogen?

The answer was yes. A few hours before the explosion, the 2-in diameter open vent on the tank had been replaced by a filter pot containing alumina, presumably to prevent moist air contaminating the MEK in the tank. The pressure drop through the alumina was sufficient to allow the pressure in the tank, designed for an 8-in water gauge (2 kPa or 0.3 psi), to be exceeded and to reach the rupture pressure, probably about a 24-in water gauge (6 kPa or 0.9 psi). The nitrogen flow rate was estimated to be $\approx 11 \, m^3/min$ (400 ft^3/min) so the rupture pressure could be reached in about an hour. Unfortunately the report does not say how much time elapsed between the start of blowing and the rupture. The fire that followed the explosion could have been the result of sparks produced by the tearing of the roof-to-wall joint.

Once again we see a change made without adequate consideration given to the possible consequences. In this case, the result was inevitable and occurred soon after the change. In the previous case history, the result was probabilistic and did not occur for many years.

8.3 ONE + ONE = MORE THAN TWO

We are familiar with synergy: two (or more) drugs or parts or the body work together to produce a greater effect than the sum of their individual parts. The same is true of hazards, as the following example shows:

The report [4] starts with the words, "It was the best of times; it was the worst of times. The economy was booming; some of the booms were due to plant explosions." One occurred in a power station boiler in a car factory in February 1999. The primary fuel was pulverized coal but natural gas was also used. There were two gas supply lines, each of which supplied three burners. The boiler was shutting down for overhaul. One of the natural gas lines was isolated and blinded; the valves between the blind and the burners were opened and the line swept out with nitrogen. The other line had not yet been blinded. In addition, the valves in this line had been opened in error (or perhaps left open). Gas entered the furnace. There were no flame-sensing interlocks to keep the inlet motor valves closed when there was no flame and after 1.5 min, an explosion occurred. The ignition source was probably hot ash. The explosion inside the boiler set off a secondary explosion of coal dust in the boiler building and in neighboring buildings. Six employees were killed and many injured. Damage was estimated at one billion dollars, making it the most expensive industrial accident in US history.

There were thick accumulations of coal dust in the damaged buildings. Even after the explosions, the dust was an inch thick. On many occasions, a primary explosion has disturbed accumulations of dust and resulted in a far more damaging secondary explosion. The hazards of dust explosions and the need to prevent accumulation of dust are well established. Henry Ford is reputed to have said that history was bunk. Did the car factory personnel still believe Ford's words in 1999?

The same paper also describes another furnace explosion, killing three employees, also in February 1999. There seems to have been a flame-out and, as in the first incident, there were no flame-sensing interlocks to close the fuel gas motor valves when flames went out. In addition, one fuel gas valve was leaking. As in the first incident, the primary explosion disturbed dust, resin this time, and caused a secondary explosion.

8.4 "NEAR ENOUGH IS GOOD ENOUGH"

Anyone who has bought a new house (at least, in the UK) knows how difficult it can be to get the builders to finish everything and the new owners often move in when a few jobs are still outstanding. It may not matter if the builders have not finished laying the paths or fitting out the guest room, but would you start up plant equipment before it was complete? Here is the story of a company that did, perhaps because it was just a storage tank, not a production plant [5].

Three low-pressure storage tanks were being modified for the storage of crude sulfate turpentine, an impure recovered turpentine with an unpleasant smell and a flashpoint that can be as low as 24 °C (75 °F). Several changes were being made:

- A fixed foam fire-fighting system was being installed, with a pumper connection outside the dike.
- To prevent the smell reaching nearby houses, a carbon bed would absorb any vapors in the vents.
- Flame arresters in the common vent system would prevent an explosion in one tank igniting the vapor in the others.

Movement of turpentine into the tanks was started six weeks before all the protective equipment could be fitted. Each tank contained $\approx 800\,m^3$ (200,000 gal). The local authority who gave permission for the storage was informed and seems to have raised no objection. Twelve weeks later the protective equipment was still not complete but the vent absorption system was ready and was brought into use. All three tanks were connected to the vent absorption system and no longer vented directly to the atmosphere.

The manufacturer's instructions said the carbon bed had to be kept wet. It was not and got too hot. During the day, the oxygen content in the tank was too low for ignition but rose in the evening when the tank cooled and air was sucked in. The hot carbon ignited the vapor and there was an explosion. It spread to the other two tanks through the common vent collection system, as the flame arresters had not been delivered.

The fixed foam firefighting system on the tanks could not be used as the piping connection outside the bund had not been installed.

The explosion damaged three other tanks in the same dike. One contained an acidic liquid and another an alkaline one. The acid and alkali reacted and produced hydrogen sulfide. Incompatible liquids should not be stored in the same dike.

The vent absorption system was intended to prevent pollution. Because it was operated incorrectly, because the missing flame arresters allowed the explosion to spread, and because the firefighting equipment was incomplete, the result was an environmental disaster. Two thousand people were evacuated from their homes for several days and 10–15 hectares (25–40 acres) of marsh were contaminated.

Note that before the carbon bed was commissioned, an explosion was possible but unlikely. Commissioning it before the rest of the new safety features were ready and not keeping it wet made an explosion inevitable.

What, I wonder, were the qualifications, abilities, knowledge, and experience of the people in charge of the plant involved in this incident? What pressure, I wonder, was put on them to bring the tank into use prematurely? Near enough may not be good enough.

8.5 ANOTHER EXPLOSION IGNITED BY A CARBON BED

A carbon absorption bed was added to the vent system of an ethylbenzene tank to absorb vapor emissions. It was designed to handle only the emissions caused by changes in the temperature of the tank. The much larger emissions produced when the tank was being filled were sent to a flare stack.

One day, when the tank was being filled, the operator forgot to direct the vent gases to the flare stack. When the tank was a 25% full, he remembered and promptly corrected the error. This is understandable. When we realize that we have forgotten to carry out a task, we tend to do it at once, without stopping to ask if there might be any adverse result. When the filling was complete, he sent the emissions back to the carbon bed. Within minutes the carbon bed caught fire; damage was severe.

When the carbon bed received far more vapor than it was designed to absorb, it overheated. When the vapor was sent to the flare stack, the carbon bed and the absorbed vapor could not burn as there was no air (or not enough air) present. When filling was complete and the tank was again connected to the carbon bed, it received a supply of air from the vent and was still hot enough to ignite the absorbed ethylbenzene and then the carbon.

As already stated, the operator's error was understandable (see also Section 13.1). However, during the design of the system someone should have asked what would occur if the vent stream was wrongly directed. A Hazop would have raised this question. After the fire various protective devices were considered:

- An interlock to prevent vapor being directed to the carbon bed while the tank was being filled.
- A high-temperature alarm on the carbon bed.
- A carbon monoxide detector on the carbon bed to detect smoldering.
- Nitrogen blanketing of the tank.

The report [6] does not say what was actually done but the last proposal is the best as it will prevent explosions from all sources of ignition.

Many other fires and explosions have occurred in vent collection systems, installed without sufficient thought, for the commendable purpose of improving the environment (see *WWW*, Section 2.11). Two more follow. Under the section, "Green Intention, Red Result," Reference [7] describes these and other changes which were made to improve the environment but had adverse effects on safety (see also Section 8.9).

8.6 AN EXPLOSION IN AN ALTERNATIVE TO A CARBON BED

Alternative methods of removing volatile and flammable contaminants from a stream of air are to burn them in a furnace or oxidize them over a catalyst. The concentration of vapor is kept below the lower flammable limit (LEL) to avoid an explosion. The concentration is measured continuously and if it approaches the LEL, the operation is automatically shut down.

A trip on a new oxidizer kept operating. A check with a portable combustible gas detector showed that the plant instrument was reading high. The startup team therefore decided to take the trip system out of use while the reason for its high reading was investigated but to continue with the startup without it.

Many people have taken a chance like this and gotten away with it. The team on this unit was not so lucky. Within 2 hr, there was an explosion with flying debris. It is not clear from the report [8] whether or not the plant instrument was really reading high, but it is clear that there were occasional peaks in the vapor concentration.

8.7 ONLY A MINOR CHANGE

A reactor vent discharge containing 100 ppm benzene in nitrogen was sent directly to the atmosphere at a rate of $8.5\,m^3$/hr ($5\,ft^3$/min). To meet new emission standards, the company installed an electric flameless destruction system. The vent discharge had to be diluted with air before entering this system and the air rate was set so that the total flow was $170\,m^3$/hr ($100\,ft^3$/min). This dilution ensured that the mixture was well below the lower flammable limit of benzene even during occasional spikes when the benzene concentration rose briefly to 15%.

Shortly after installation of the destruction unit, the vent discharge from a storage tank was also directed into it. The increase in flow rate was only 6.7%. Everyone assumed that this was too small to matter and no one made any calculations. However, the lower flammability limit was exceeded during the spikes in benzene concentrations in the main contributor to the flow. The destruction unit was hot enough to ignite the vapors and there was an explosion. A high concentration of combustible gas in the gas stream sounded an alarm but it operated too late to prevent the explosion. Although damage was considerable, the explosion did not travel back to the reactor and tank as both were blanketed with nitrogen.

8.7.1 Lessons Learned

Consider the possible consequences of changes before authorizing them (see Chapter 2). Never dismiss a change in quantity as negligible before calculating its effects. Consider transient and abnormal conditions as well as normal operation. Sections 3.1.2, 4, and 5 describe other incidents that occurred because no one made simple calculations.

Estimate the response time of every alarm and trip to see if it is adequate. Check it during testing if there is significant delay. Most measuring instruments respond quickly but analytical instruments are often slow, although it is usually the sampling system rather than the measuring device that causes the delay.

The report [6] says that pollution control equipment should not be treated like a domestic garbage can, something into which anything can be dumped. Every proposed addition should be thoroughly evaluated. On a chemical plant or in a chemical laboratory this applies to all waste collection equipment. Many fires, toxic releases, or rises in pressure have occurred because incompatible chemicals were mixed in the same waste drum (see Section 8.11).

8.8 AN EXPLOSION IN A PIPE

The pipe (C) (Figure 8-1) transferred fractionation residues from a batch distillation vessel (A) to residue storage tank (D) via the reversible pump (B). Distillation residues from other units and condensate from vent headers also went into tank D. When D was full, the contents were moved to A for fractionation and recovery. As the residues were viscous, pipe C was steam-traced.

This part of the plant operated only five days per week. It was left one Friday evening after the contents of D had been moved to A, ready for distillation on Monday. Over the weekend, a discharge reaction ruptured pipe C.

Analysis of the remaining material in other parts of pipe C showed that decomposition and self-heating started at $\approx 140\,°C$ (280 °F) and that the rate of temperature rise soon exceeded 1,000 degrees C (1,800 degrees F) per minute. This was surprising as the residues reached 140 °C in normal operation and had never shown signs of decomposition or exothermic activity. Further investigation showed that the instability was due to the presence of 3% water that had entered vessel D with the condensate from the vent headers. Water is a very reactive substance and can form unstable mixtures with many other compounds. The disaster at Bhopal, India was due to the contamination of methyl isocyanate with water.

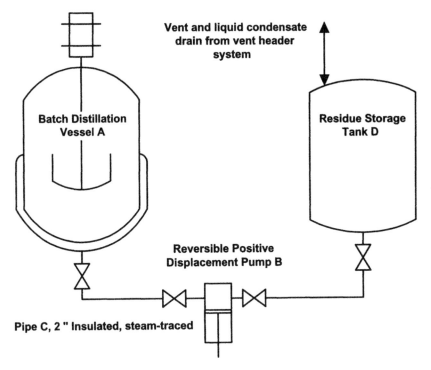

Figure 8-1. Some of the residue moved from D to A and left in pipe C decomposed and ruptured the pipe. From reference 9. Reprinted with the permission of the American Institute of Chemical Engineers.

As almost always, something else was also wrong. The steam supply to the tracing on line C came from an 8.3 bar gauge (120 psig) supply via a let-down valve, which had failed in the open position. This raised the temperature of the pipe to 170 °C (340 °F), high enough for decomposition to start.

8.8.1 Lessons Learned

A relief valve was fitted downstream of the steam let-down valve. An alternative and inherently safer solution would have been to use a heating medium that could not rise above 140 °C (280 °F).

Because water is so reactive and present most everywhere, we should, during Hazop studies, ask, under the heading *Other than*, if water could be present and, if so, what its effects would be. (A Hazop was carried out elsewhere on an existing plant in which some valves were operated by high-pressure compressed gas. The team was asked if water could be present in the gas and the members all agreed that this was impossible. None of them knew that during shutdowns, when no high-pressure gas was available, the maintenance team occasionally used high-pressure water to operate the valves. See also Section 11.1.7) We should also always ask if other common contaminants such as rust, lubricating oil, and any material used elsewhere in the process (see Section 8.1.1) could be present.

The vent line drains did not originally go to tank D but were diverted there to reduce waste. Perhaps because this was obviously a good deed, its possible consequences were not thought through. Chapter 3 describes other changes that had unforeseen results. As the report [9] says, "No good deed goes unpunished."

As shown by many of the other incidents described in this book (e.g., Sections 3.2.7 and 8.1), a plant can operate for many years without incident until a slight change in conditions results in an accident.

8.9 A DUST EXPLOSION IN A DUCT

The exhaust stream from a dryer which contained volatile organic compounds and some flammable dust was discharged to the atmosphere through a vertical vent stack. To comply with legislation, the vent stack was replaced by an incinerator. There was no room for it near the dryer so it was built 90 m (300 ft) away and connected by a long duct. The dust settled out in the duct and was removed every 6 mo, by which time it was 3–25 mm ($\frac{1}{8}$–1 in) thick.

The dryer was shut down for maintenance but the incinerator was left on line. When the dryer was brought back on line there was an explosion, which killed one man and caused extensive damage. The probable cause was a pressure pulse from the startup of the dryer, which disturbed enough of the dust in the duct to produce a small explosion, which then disturbed and ignited much of the remaining dust in the duct. A layer of dust <1 mm thick can, if disturbed, produce an explosion in a building.

There should have been an explosion detection-and-suppression system or explosion vents in the duct. Or better still, filters to remove the dust before it entered the duct [6].

Once again we see with this example that a change meant to reduce pollution was made as cheaply as possible and without adequate consideration of the hazards. It seems that when people are faced with an environmental problem, a sort of tunnel vision can set in and all thoughts of side effects are brushed aside (see Sections 8.5–8.7).

8.10 OBVIOUS PRECAUTIONS NEGLECTED

An underground concrete tank, 27 m (88 ft) diameter, 4 m (13 ft) tall, capacity 720 m^3 (190,000 gal), had been out of use for several years. It had a concrete roof supported by 27 internal columns and covered by a meter of soil. It was decided to recondition the tank. Two holes were cut in the roof of the tank for the insertion of new instruments. Before work started, the concentration of flammable vapor in the tank was checked and found to be <1% of the lower flammable limit.

During the weekend no work was carried out on the tank but several loads of product arrived by barge and were transferred into neighboring tanks. The last was a load of premium gasoline. It was followed by a water flush, directed at first into the gasoline tank and then after 10 min into the concrete tank.

On Monday morning, three welders started work again. No flammability tests were carried out. When the first torch was lit, the tank exploded. The three welders were blown off the top of the tank and killed. Soil was thrown almost 100 m (325 ft).

Calculations showed that a gauge pressure of at least 0.43 bar (6.2 psi) would have been needed to lift the roof off its supports. The tank was thus much stronger than the usual atmospheric pressure storage tank, which will rupture when the gauge pressure in it exceeds 0.06 bar (0.9 psi) (see Section 8.2). The explosion of as little as 0.7 m^3 (180 US gal) of gasoline could have been developed sufficient pressure to lift the roof off its supports.

8.10.1 What Went Wrong?

- The inlet line to the tank should have been blinded before welding started in order to prevent anything from leaking through it while the transfer line was in use.

- Even it there had been no movements over the weekend, the atmosphere in the tank should have been checked before work was resumed that Monday. A test on Friday (or earlier) does not prove that equipment is still safe on Monday.

- It was naïve to assume that no gasoline would be left in the transfer pipe after flushing with water for only 10 min.

- Did the owners leave testing to the reconditioning contractors? Did they know what had occurred during the weekend? The report [10] does not say.

The design of the tank made it stronger than usual; when it did fail, it failed with greater violence. Stronger does not always mean safer.

8.11 A DRUM EXPLOSION

This was a small explosion and fire compared to those described in the preceding, and no one was injured, but it was investigated with commendable thoroughness. The 210-l (55-gal) drum contained a peroxide (*di-t*-butyl peroxide). It was kept in a horizontal position on a cradle and small amounts were withdrawn as needed through a small cock on the lid, weighed, and added to a batch reactor. The weigh station was also used for other materials. The explosion blew the lid off the drum. It landed 15 m (50 ft) away with its outside surface on the ground but with soot on this surface. The location of the soot, and its nature, indicated that the lid had been exposed to fire before the explosion and that the heat from this fire caused explosive decomposition of the peroxide.

The location of the initial fire was either a drip tray underneath the cock or a cardboard box containing flammable materials located under the weigh table. The fire that followed the explosion caused most of the damage. No source of ignition was found but peroxides are very easily ignited. It often happens, after a fire or explosion, that the source of ignition is never found.

The report [11] recommended that the peroxide should be supplied in 20-l (5-gal) containers in future so as to reduce the inventory of this unstable substance ("What you don't have can't decompose") and that housekeeping should be improved. Although contamination of the peroxide was ruled out on this occasion, it was possible for it to occur. It was decided to use a dedicated weigh station in future.

Often during the investigation of an accident, several scenarios are considered possible but on the balance of evidence, one is considered more likely than the others. If the others could have occurred, as was the case here, then we should take actions to prevent all possible/likely causes in the future.

Other drums have exploded (or bulged) because they were used for waste materials that reacted with each other. People have been injured when removing the lids from bulging drums as the lid flies off as soon as the closing mechanism is released. Empty drums have exploded because vapor from the previous contents was still inside. New drums may contain traces of solvents used by the manufacturers to clean them. Never use drums as access platforms, especially for hot work.

8.12 FOAM-OVER — THE CINDERELLA OF THE OIL AND CHEMICAL INDUSTRIES

I have often drawn attention to the way the same accidents keep recurring, sometimes in the same company, despite the publicity they get at the time [12]. Unfortunately, after perhaps 10 yr, most of the people at a plant have left or moved to another department, taking their memories with them. Their successors do not know the reasons for some of the procedures introduced after an accident and, keen to improve output or efficiency, both very desirable things to do, make a number of changes. The accident then happens again.

One of these accidents that keeps recurring, despite frequent publicity, is a foam-over or slop-over. It occurs when hot oil, over 100 °C (212 °F), is added to a tank containing a water layer, or oil above a water layer is heated above 100 °C. The heat travels down to the water, which then vaporizes with explosive violence, often lifting the roof off the tank and spreading the oil over the surrounding area. The oils involved are usually heavy oils or tars, which have to be heated before they can be pumped, and they cover everything with a thick black coat.

Waste liquids were distilled to remove water and light ends and the residue was used as fuel. It was stored in a vertical cylindrical tank ≈12 m (40 ft) tall and 3.6 m (12 ft) diameter, volume ≈120 m^3 (30,000 gal). The bottom meter of the tank was conical. As the result of a plant upset, some water got into the tank. When hot oil was being run into the tank, the roof parted company with the walls and about 40 m^3 (10,000 gal) of hot black oil was blown out.

The tank had been filled without incident ten times since the plant upset. A solid crust probably insulated the water in the conical bottom section of the tank until something caused it to move or crack. Calculations showed that as little as 30 kg (65 lb) of water could have produced enough steam to produce the damage that occurred.

To prevent foam-overs, if heavy oil is being moved into a vessel that may contain water, the temperature of the oil should be kept <100°C (212°F) and a high-temperature alarm should be fitted to the oil line. Alternatively,

- Drain the water from the tank.
- Keep the tank >100°C to evaporate any water that leaks in.
- Circulate or agitate the contents of the tank before starting the movement.
- And start the movement at a low rate.

When heavy oil is moved out of a vessel, drawing it from the bottom will prevent small amounts of water from accumulating.

This accident illustrates another feature of many industrial accidents: An operation can be carried out many times before a slight variation in conditions results in an accident. A blind man can walk along the edge of a cliff for some distance without falling but that does not make it a safe thing to do.

The report [13] on this foam-over does not draw attention to the fact that there have been many similar incidents in the past and many published accounts of them, for example, *Hazards of Water* [14], published in 1955, contains many accounts of tanks and pressure vessels damaged by the sudden vaporization of water. They are also described in *WWW*, Section 12.2. Why then do they keep occurring? Perhaps people ignore reports of past accidents in the belief that the lessons must surely have been learned and incorporated in instructions and codes of practice. But the reasons for them are often forgotten or ignored. And they can never cover every possibility. They can never prohibit every possible action we should NOT take. The best prophylactic is knowledge of the hazards (see also Sections 2.5.2, 7.4, 7.5, and 14.5).

REFERENCES

1. Anon. (1992). The largest explosion in the UK, Fauld 1944. *Loss Prevention Bulletin*, Feb., 103:17–18.

2. Riezel, Y. (2002). Explosion and fire in a gas-oil fixed roof storage tank. *Process Safety Progress*, **21**(1):67–73.

3. Searson, A.H. (1983). Explosion in cone-roof gas-oil tank. *Proceedings of 4th International Conference on Loss Prevention and Safety Promotion in the Process Industries*, Institution of Chemical Engineers, Symposium Series No. 80, Institution of Chemical Engineers, Rugby, UK.

4. Zalosh, R. (2000). A tale of two explosions. *Proceedings of the AIChE Annual Loss Prevention Symposium*, March.

5. Chung, D. (2000). Explosions and fire at Powell Duffryn Terminals, Savannah, Georgia. *Proceedings of the AIChE Annual Loss Prevention Symposium*, March.

6. Myers, T.J., H.K. Kytömaa, and R.J. Martin (2002). Fires and explosions in vapor control systems. *Proceedings of the AIChE Annual Loss Prevention Symposium*, March.

7. Kletz, T.A. (2001). *Learning from Accidents*, 3rd ed., Butterworth-Heinemann, Oxford, UK and Woburn, MA, Chapter 26.

8. Baker, R. and A. Ness (1999). Designing and operating thermal oxidisers. *Loss Prevention Bulletin*, 146:8.

9. Hendershot, D.C., A.G. Keiter, J. Kacmar, J.W. Magee, P.C. Morton, and W. Duncan (2003). Connections: How a pipe failure resulted in resizing vessel emergency relief systems. *Process Safety Progress*, 22(1):48–56.

10. Baker, Q.A., D.E. Ketchum, and K.H. Turnbull (2000). Storage tank explosion investigation. *Proceedings of the AIChE Annual Loss Prevention Symposium*, March.

11. Antrim, R.F. *et al.* (1998). Peroxide drum explosion and fire. *Process Safety Progress*, 17(3):225–231.

12. Kletz, T.A. (1993). *Lessons from Disaster — How Organisations have No Memory and Accidents Recur*, Institution of Chemical Engineers, Rugby, UK.

13. Ogle, R.A. (1998). Investigation of a steam explosion in a petroleum product storage tank. *Process Safety Progress*, 17(3):171–175.

14. Anon. (1955). *Hazards of Water in Refinery Systems*, American Oil Company, Chicago, IL, 1st ed.

Chapter 9

Poor Communication

"Where is the tea? I can't find it." "It's in the cocoa tin marked 'Coffee.' "

— E. Esar, *The Humor of Humor*

The whole of this book is about poor communication. When an accident occurs in the process industries, outsiders might think that it happened because no one knew how to prevent it. While the people at the plant at the time, or the designers, may not have known, the information is almost always available somewhere. Very few accidents occur because no one knew that there was a hazard. Sometimes the behavior of a compound or reaction takes everyone by surprise but in most cases this is the result of inadequate testing, the need for which is well known. In this chapter, we look at communication in a narrower sense.

9.1 WHAT IS MEANT BY *SIMILAR*?

Some changes had to be made to a length of low-pressure ventilation ductwork ≈0.6 m (24 in) diameter. To keep the rest of it in operation during the modification, a bypass of almost the same diameter was made around the affected section. To isolate this section, the contractor was told to drill a hole in the main duct and push an inflatable rubber bladder through it. This is a standard item of equipment that had been used successfully on previous occasions. The drawing specified "[manufacturer's name] inflatable pipeline stopper or similar." This manufacturer's stopper is fitted with a metal inflation tube that ensures that the balloon remains in position beneath the insertion hole. The contractor instead used a balloon fitted with a flexible tube. The inflated balloon moved a little way down the duct and blocked the bypass line (see Figure 9-1). Ventilation flow was stopped. The operators in the control room had been warned that changing over to the bypass line might cause the low flow alarm to operate and therefore they ignored it. Some time elapsed before they realized what had happened.

The immediate cause of the incident was therefore the use of the word *similar*. What seems similar to one person seems dissimilar to another. (To some people bats are similar to birds; to others a bat — fledermaus in German — is more like a flying mouse.) The word *similar* should never be used in specifications or instructions.

Another word that should not be used is *all*. If someone is asked to remove all the slip-plates from a tank, or lubricate all the machines in a unit, he or she does not know whether there are 2, 3, 4, or many. Each one should be specified by name or number (see Section 2.1). Other words

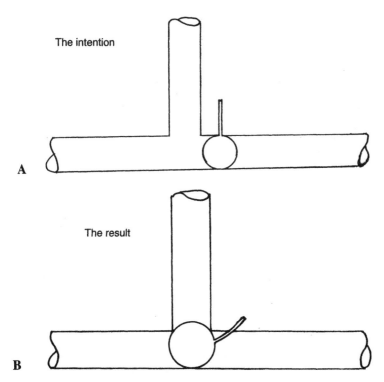

Figure 9-1. An inflatable stopper with a rigid stem was wanted but a *similar* one was fitted instead.

that should not be used are adjectives such as *large, small, long,* or *short* (see Section 3.1.5).

During an incident investigation, it is often useful to ask why the incident happened when it did and not at some other time? Asking this question disclosed another cause in the incident under discussion.

The process foreman would normally have taken a close interest in the job. He might have spotted that the wrong sort of balloon was being used and he would certainly have been better able than the operators to handle the plant upset that occurred when the ventilation flow stopped. However, he was busy with other changes being made elsewhere at the plant. Why were two major changes being made at the same time? They had both been requested by the regulators who had, as usual, agreed on a timescale with the company, a necessary requirement as otherwise nothing would ever get done. However, design and procurement had taken longer than expected, plant problems had caused delays and the agreed-upon date was only a few days away. If approached, the regulators would probably have agreed to a delay but the people involved may not have realized this or been reluctant to admit that they could not achieve what they had agreed to achieve. People are sometimes accused of taking chances to maintain production; on this occasion they took a chance in order to satisfy the regulators and to complete a safety job on time. The result was less safety and lost production.

The story got around, reaching the local press, which reported that the plant upset occurred because the operators had ignored an alarm. This was in a sense true but misleading. They had ignored the alarm because they had been told to do so. Do not believe rumors — or newspapers.

9.2 MORE *SIMILAR* ERRORS

A company had a thorough system of vessel inspection. Most of the vessels were inspected every few years but if a group of similar vessels were used on a similar duty, just one vessel in the group was inspected every 2 yr. If no corrosion was found, the other vessels were not inspected until their turn came around. The maximum period between inspections was 12 yr.

However, what is a *similar duty*? After an absorption tower on a nitric acid plant had leaked, it was realized that it operated at 100–125 °C (212–257 °F) while the other towers in the inspection group operated at 90 °C (194 °F). The higher temperature increased the rate of corrosion.

Similarly, a change was made to the transmission of a two-rotor helicopter. The manufacturer decided to test the new design on the aft transmission, as in the past it had developed slightly more problems than the forward transmission. The new design passed the tests but failed in service on the forward transmission. The helicopter crashed, killing 45 people [1].

9.3 WRONG MATERIAL DELIVERED

On many occasions the wrong material has been delivered. Here is an example [2].

A UK chemical company ordered a load of epichlorohydrin, a toxic and flammable chemical, from a supply agent and not from the manufacturer. A transport company collected the chemical from the manufacturer and changed the delivery note to one bearing the name of the agent. The man who did so made a slip and entered the number of the wrong tank container, one containing sodium chlorite. This container was therefore delivered to the company and offloaded into the epichlorohydrin tank. A violent reaction and explosion occurred, several people were injured, and large amounts of fume and smoke led to the closure of main roads and a major river crossing. Fines and costs amounting to $150,000 were imposed on the company for not testing the material before offloading it and on the transport company for delivering the wrong material.

There have been many similar incidents. Before accepting any process materials, companies should sample and analyze them to confirm that they are the material ordered. Some companies that used to do this stopped doing so when their suppliers were able to show that their procedures met quality standards. However, there are too many opportunities for error in the course of filling, labeling, and transporting, to justify this action, or rather inaction. There are further examples in *WWW*, Chapter 4 and Section 3.2.8 of this book while Section 6.4.1 shows that engineering materials should also be checked.

In the case described, changing the paperwork en route introduced an avoidable opportunity for error.

9.4 PACKAGED DEALS

When companies buy equipment such as boilers or refrigeration units that are sold already fitted with instruments and relief devices, they do not always check to see that the equipment complies with their usual safety standards, or even with acceptable standards, or that relief valve sizes have been estimated correctly. Here are some incidents that have occurred as a result:

- A contractor supplied nitrogen cylinders complete with a frame for holding them, a reducing valve and a hose for connecting them to the plant. The end came off the hose and injured the operator, fortunately not seriously. If anyone in the company had tried to order the hose as a separate item, it would never have made it through the purchasing procedure. An engineer would have examined the drawing and found that it did not conform to the company's standards. Furthermore, any hose acquired in this way would have been registered for a regular pressure test. But as part of a *package deal* it slipped through.

- A reciprocating compressor was started up in error with the delivery valve closed. The relief valve was too small and the packing round the cylinder rod was blown out. The compressor had been in use for ten years but the users did not know that the relief valve was merely a *sentinel* valve to warn the operator and that it was incapable of passing the full output of the compressor.

- A reciprocating pump was ordered that was capable of delivering $2\,m^3/h$. The manufacturer supplied his nearest standard size, which was capable of delivering $3\,m^3/h$, but sized the relief valve for $2\,m^3/h$. When the pump was operated against a restricted delivery, the coupling rod was bent. Fortunately, it was the weakest part of the system.

- A specialist contractor was making an under-pressure connection to a pipeline when a $1/4$-in branch was knocked off by a scaffolding plank. The company did not allow $1/4$-in connections on process lines — all branches up to the first isolation valve were 1 in minimum — but they did not check the contractor's equipment.

- The support legs on a tank trailer, used to support the tank when it was not connected to a tractor were designed in such a way that they could not be lubricated adequately. Several failures occurred.

- A relief valve, supplied with a compressor, was of an unsuitable type, and was mounted horizontally and vibrated so much that the springs dented the casing. Relief valves should be mounted vertically so that any condensation or dirt which collects in them has the maximum chance of falling out.

- Packaged equipment may not use the same threads as the main plant. This is probably a bigger problem in the UK than in the US as several different types have been in use in the UK during the lifetimes of old plants.

9.5 "DRAFTSMEN'S DELUSIONS"

Elliott [3] uses the term "draftsmen's delusions" to describe problems that occur because the beliefs of the drawing office differ from the reality of the plant. Others call them misconceptions [4].

For example, a small solvent drying unit was designed to operate at a pressure of 2 bar gauge (30 psig). The drying chambers had to be emptied frequently for regeneration so a nitrogen connection was needed. The designer looked up the plant specifications and found that the nitrogen supply operated at a pressure of 5.5 bar gauge (80 psig). This was far above the unit's operating pressure so the designer assumed there was no danger of the solvent entering the nitrogen main by reverse flow and supplied a permanent connection. (He supplied a check valve in the line but these are not 100% effective. They would be more effective if they were regularly maintained but rarely are; we cannot expect equipment containing moving parts to work forever without maintenance.)

If the designer had asked the operating staff, they would have told him that the unit was to be located near the end of the nitrogen supply line and that its pressure fell to <2 bar gauge when other units were using a lot of nitrogen. If the designer had ever worked at a plant, he would have known that it is by no means uncommon for nitrogen supply pressures to fall, especially when large units are being shut down for maintenance or are being swept out ready for start-up.

On the drying unit some solvent, which was flammable, entered the nitrogen main by reverse flow and then entered another item of equipment where it exploded [5].

If the designer had known that the nitrogen supply was unreliable, he would have fitted a low-pressure alarm to the supply and a more positive isolation on the connection to the plant (such as double-block-and-bleed valves or a hose that can be disconnected when not in use). There is more about this incident in Section 10.9.

9.6 TOO MUCH COMMUNICATION

A password had to be entered into a control computer before it was possible to override a software interlock. The monthly test of the interlock showed that unknown to the operators it had been overridden. It was then found that the password had been given, officially, to 42 people! We cannot expect every one of 42 people to keep a secret or not to misuse it.

If an interlock, trip, alarm, or any other protective device has to be overridden or taken out of use, via a computer or in any other way, this should be signaled in a clear and obvious way, for example, by a light on the panel, a note on the screen, or a prominent notice.

9.7 NO ONE TOLD THE DESIGNERS

There have been many failures of equipment because the wrong grade of steel was used (see Section 6.4.1 and *WWW*, Chapter 16) but most of them have been the result of errors by suppliers, construction teams, or maintenance teams. Here is one with a different cause.

The failure of a boiler tube in a power plant caused a steam explosion, that is, the rapid vaporization of water. It wrecked the combustion chamber and surrounding equipment. The tube failed because the grade of steel specified by the designer was unsuitable for the duty. What was worse, the same company built an identical boiler, using the same grade of steel, after the failure. It also failed. The underlying cause was not the failure of the steel but the failure of the company's internal communication system [3].

Thus we end this chapter as we began. Many companies have no formal or informal procedure for passing on information on the causes of accidents and the action needed to prevent them from happening again. In the UK, the regulators have instructed at least one major company to set up a formal system.

Commenting on the explosion at Longford, Australia in 1998 (see Section 4.2), Watkins writes [6], "The operators were quite willing to report. The problem was that the system at the time did nothing with the reports."

REFERENCES

1. Air Accidents Investigation Branch (1989). *Report on the Accident to Boeing 234LR, G-BWFC 2.5 miles east of Sumburgh, Shetland Isles on 6 November 1986*, Her Majesty's Stationery Office, London, p. 36.

2. Anon. (2000). Action following the explosion at Avonmouth on 3 Oct. 1996. *Loss Prevention Bulletin*, June, **153**:9–10.

3. Elliott, P. (1998). Overcome the challenge of corrosion. *Chemical Engineering Progress*, **94**(5):33–42.

4. Das, B.P., P.W.H. Chung, J.S. Busby, and R.E. Hibbered (2001). Developing a database to alleviate the presence of mutual misconceptions between designers and operators of process plants, *Hazards XVI: Analysing the Past, Planning the Future*, Symposium Series No. 148, Institution of Chemical Engineers, Rugby, UK, pp. 643–654.

5. Kletz, T.A. (2001). *Learning from Accidents*, 3rd ed., Butterworth-Heinemann, Oxford UK and Woburn, MA, Chapter 2.

6. Hopkins, A. (2002). Lessons from Longford — The Trial, *Journal of Occupational Health and Safety — Australia and New Zealand*, Dec. **118**(6):1–72.

Chapter 10

I Did Not Know That . . .

The recipe for perpetual ignorance is: be satisfied with your opinions and content with your beliefs.

— Elbert Hubbard (1859–1915)

This chapter describes some incidents that occurred because some of the properties of the materials and equipment used were unknown to those who handled them.

10.1 . . . THAT METALS CAN BURN

Thin metal packing was increasingly used during the 1980s and this change was followed by an increase in metal fires. Many people did not realize that metals burn quite readily when they are in the form of powders or thin sheets and can produce higher temperatures than oil fires. Aluminum and iron, not normally considered flammable, as well as titanium and zirconium, can burn when in these forms and the fires are difficult to extinguish. Small amounts of water may be decomposed into hydrogen and oxygen and can worsen the fire. Water should not be used for firefighting unless a large quantity is available to quickly drench a very small fire. Burning can continue in atmospheres of carbon dioxide, nitrogen, and steam, and the burning metal can react vigorously with other materials. Argon can be used for firefighting and special agents are also available [1,2]. If any metal oxides are present, a hot metal with a greater affinity for oxygen can react with it (the thermite reaction). For example, hot aluminum or titanium can react with the oxygen in rust and produce enough heat for a self-sustaining reaction.

Bulk metal is, of course, more difficult to ignite but several titanium heat exchangers, including their tube sheets, have been destroyed by self-sustaining fires. A strong ignition source, such as a welding torch, is needed but it need not impinge directly on the titanium. The hot slag formed by cutting steel contains iron oxide and can start a thermite reaction [3].

A fire occurred in a packed column, ≈75 m (250 ft) tall and 8 m (25 ft) diameter, packed with carbon steel. It started when the packing and column internals were being removed; the fire was ignited by hotwork. The fuel was the steel packing, possibly supplemented by process materials that had not been completely removed even though the column had been steamed. Because of the high surface area of packing, it is always difficult to be sure that it is completely clean. Even new packing may be coated with oil.

Hot work should be avoided, if possible, above or below packed beds. If it cannot be avoided, for example, by removing the packing first, then the possibility of a fire should be considered and a plan for dealing with it prepared [4].

Increasing the thickness of metal packing makes ignition more difficult, but this increases the heat produced if the packing does ignite. Decreasing the spacing between the metal sheets also makes it harder to ignite them but they are then more likely to become contaminated by process liquids and harder to clean. Trade-offs have to be made between these factors and the weight and efficiency of the packing.

10.1.1 Another Metal Fire

This fire occurred in a column 22 m (73 ft) tall and a ≈1 m (40 in) diameter, which contained titanium packing. The performance of the column showed evidence of plugging so it was taken out of service and prepared for entry. Small pieces of titanium were observed on the redistribution tray above the middle of the three beds.

A flash fire occurred in the packing — perhaps not all the process material had been removed — and a few minutes later a bright spot of glowing metal was noticed. It grew rapidly in size and destroyed a whole section of the packing. The most likely source of ignition was pyrophoric deposits and the fire may have started in the small pieces of titanium. While titanium in bulk self-ignites at 1,120 °C (2,050 °F), powdered titanium ignites at 330 °C (625 °F). It is not known whether the titanium fire started before or after the flash fire [5] (see also Section 6.1.6).

10.2 . . . THAT ALUMINUM IS DANGEROUS WHEN WET

There were several tank trucks and trailers on the lower deck of a roll-on, roll-off ferry when a smell of ammonia was detected on the lower deck. There were several loads of dangerous goods on board but none of the vehicle drivers could explain the smell. It seemed to come from a trailer carrying metal products. Its sides were hot and water was dripping out of the bottom.

The trailer contained aluminum waste and turnings, which can produce hydrogen when wet. This has been known for many years and Bretherick quotes reports dating back to 1947, including a patent for a propellant explosive made from aluminum and water [6]. It is difficult to see how ammonia could be formed. According to the International Maritime Dangerous Goods Code, the load should have been classified as "dangerous when wet" (that is, it can produce a flammable gas on contact with water and sufficient energy to ignite it) and should be packed and labeled accordingly. Ventilation was increased and the ship reached its destination without incident [7].

10.3 . . . THAT RUBBER AND PLASTICS ARE PERMEABLE

A former colleague of mine has described a New Year Ball that did not go exactly as planned [8].

We wished to make the New Year Ball particularly spectacular and had arranged for a couple of hundred brightly colored balloons to be released among the revelers from a net suspended from the ballroom ceiling. During the afternoon before the event, we decided that manual inflation of the balloons was far too exhausting and I ordered a cylinder of compressed carbon dioxide to be sent up from the Analytical Lab. The balloons, all two hundred of them, were inflated in no time at all and the clusters were hoisted to the ceiling in the releasable net. Imagine our chagrin and extreme embarrassment when, upon arriving for the opening of the Ball a few hours later, we found that every balloon had shrunk to the size of a small orange and on eventual release fell to the floor with sickening thuds. I had learned my lesson — India rubber is permeable to carbon dioxide!

When plastic water pipes are run through oil-soaked ground, the water may become contaminated with oil (see also Section 12.2).

In some combustible gas detectors, the sample is drawn through a plastic tube to the measuring element. The plastics used absorb some flammable vapors. It is better to use detectors in which the element is at the end of a lead and can be located at the test point such as the inside of a vessel.

Plastic containers used to collect samples of gas for analysis may absorb some constituents of the gas and make the analysis results incorrect.

10.4 . . . THAT SOME PLASTICS CAN ABSORB PROCESS MATERIALS AND SWELL

In the early days of nitroglycerine (NG) manufacture, there were many explosions. These became less frequent and less damaging as the size of reactors was reduced and ultimately the original batch reactors holding approximately a ton of material gave way to continuous reactors holding perhaps a kilogram. Similar reductions in size were made to the washing and separation stages. I have often quoted these changes as an example of intensification and inherently safer design. Safer, yes, but not safe [9].

The NG was separated from surplus acid in a centrifuge (Figure 10-1). The NG caused the plastic pipe to swell so that some of the NG passed down the acid line into the acid tank and settled on top of the acid. Two explosions occurred, one in the acid tank and one in the recycle line out of the tank. Vibration probably triggered the first explosion and the sun's heat probably triggered the second.

A Hazop could have prevented the explosions, provided the team realized that "Less of" flow could occur in the NG line.

10.5 . . . WHAT LAY UNDERNEATH

Apart from ignorance of the properties of materials, many people are unaware of the way some equipment, particularly old equipment, is constructed. A small tank, capacity $\approx 100\,m^3$, held 57 tons of a liquid similar to gasoline in its physical properties. All the lines leading to it were disconnected and blanked except for one line in which the two valves were locked off. Nevertheless, in the course of 24h the level fell to 50 tons. Dipping confirmed that the fall was real and not an instrument error. No

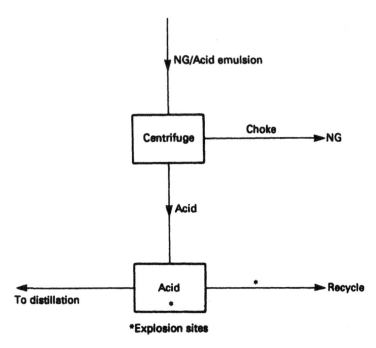

Figure 10-1. A choke caused NG to enter the acid line. A Hazop could have foreseen this.

sign of a leak could be seen even though the tank was sitting on a concrete base in a concrete-lined dike. The tank was emptied and filled with water. Again the level fell.

The drawings were found. To everyone's surprise they showed that the concrete base was only a concrete ring and that the inside of the ring was filled with sand. Holes were dug round the tank, down to the water table, but no oil was detected. There had been a lot of rain and the oil had been washed away.

The tank was lifted off its base and the sand replaced by concrete.

10.6 . . . THE METHOD OF CONSTRUCTION

An unusual method of construction produced another hidden hazard. A steel fractionation column was fitted with an internal condenser with an aluminum-bronze tubesheet. It had the same diameter as the vessel but was welded to it in an unusual way, as shown in Figure 10-2. One of the

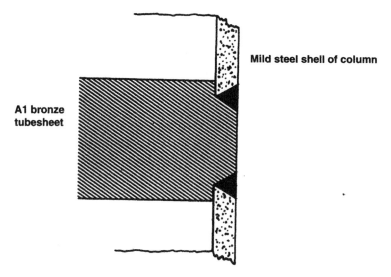

Figure 10-2. No one at the plant was aware of this unusual method of construction.

welds cracked in service and there was an escape of flammable vapor. Fortunately, it did not ignite.

The column had been inspected twice in the 5 yr since it had been built but nothing unusual was found. It is quite possible that no special attention was paid to the bimetallic welds as all the engineers there when the unit was built had left and none of their replacements now knew that the construction was unusual.

The underlying cause, of course, was the lack of any system for keeping necessary information extant. Unusual design features and points to watch during inspections should be recorded on vessel registration records and vessel inspection schedules.

Anther incident occurred when two laboratory workers, wearing air masks, were attaching a 2-l cylinder of ethylene oxide to some equipment. One of them removed part of the cylinder valve, thinking it was a protective cap. There was an escape of ethylene oxide, which was carried into the ventilation ducting and set off a gas detector alarm. The building was evacuated and the emergency team removed the cylinder and immersed it in water.

At one time an incident such as this would have been put down to operator error. There were errors but mainly by other people. According to the report:

- The laboratory workers had not been adequately trained and their knowledge had not been assessed.
- Some of the instructions were in English rather than in the language of the country where the incident occurred.
- The part that was removed should have been labeled "Do not remove."

Although the alarm system functioned correctly and the correct emergency action was taken, the investigation found that the gas detector was set at maximum sensitivity and often sounded when normal laboratory operations were carried out nearby. This could lead to the alarm being ignored and its setpoint was raised [10].

10.7 . . . MUCH ABOUT STATIC ELECTRICITY

Static electricity is a common cause of ignition but many people are not clear about the conditions necessary for it to ignite a flammable mixture. Reports on fires or explosions sometimes quote it as the source of ignition without making it clear exactly how it arose. In the following incident, the people concerned were unsure about the precautions necessary and, in addition, did not realize that the equipment they were using was unsuitable.

A batch plant contained a number of reactors and a number of small storage tanks. Because the spaghetti bowl of fixed piping needed to connect every tank to every reactor would be complex and provide many opportunities for errors and contamination, the plant instead used suction and delivery hoses and a metering pump. There is much to be said for this system but it introduces different hazards. Hoses are more easily damaged than fixed piping and can be attacked by some process materials. To prevent this, the company specified high-quality hoses, reinforced by metal coils embedded in the plastic, and suitable for all the materials handled.

The metal coils in the hoses were not connected to the end-pieces and formed isolated conductors. When a hose became worn, the ends of the spirals protruded into the interior of the hose. The flow of liquid through the hose generated a static charge and an induced charge on the coil; this charge could not flow to ground. Sparks passed between the end of a coil and a metal end-piece, which was connected to the plant and therefore grounded. Although most of the time this did not matter, as most of the liquids handled were nonflammable, one process used toluene as a raw material. In this process, a spark could pass through the liquid without igniting it but once the liquid was displaced by air, a flammable mixture would be formed and an explosion in the vapor space of the inlet vessel was the result. Fortunately, the toluene concentration was near the upper flammable limit and the explosion was not very violent [11].

It was unlikely that everyone was ignorant of all the following points but ignorance was certainly widespread:

- The people who specified the hoses did not ask for the coils to be connected to the end-pieces, either because they did not see the need for this or perhaps because they did not foresee that flammable liquids might be handled.

- The operating staff did not check that the coils were grounded before they used them with a flammable liquid.

- The usual method of checking that a coil is grounded is to measure the conductivity between the two end-pieces. However, the hoses contained three separate coils. Even if they were originally connected to the end-pieces, this test could not detect a failure of the connections on one coil as the others would carry the current. Hoses with only a single reinforcing coil should be used (or hoses with external coils so that the each coil could be checked). Three internal coils in a hose may make it mechanically stronger but there is no easy way of testing their integrity.

- It is not good practice to displace a highly flammable liquid like toluene with air.

Toluene has a low conductivity and any static charge it acquires will drain away only slowly. Conducting liquids lose their charge quickly if their container is grounded. However, this incident could have occurred with any highly flammable liquid as nongrounded metal acquired an induced charge.

Toluene was moved through the hoses many times before a hose protruded close enough to an end-piece for a spark and ignition to occur. This is typical of many accidents. *We have done it this way a hundred times* does not prove an action is safe — unless an accident on the 101st occasion is acceptable.

10.7.1 Another Static Ignition

As already stated, static electricity is often quoted as the source of igni-tion although it is not clear exactly how it arose. In the following report [12], the investigators went to considerable trouble to establish exactly what probably happened.

Some material was to be added via a hose to an intermediate size bulk container made from polyethylene and surrounded by a metal cage. There was some water and some highly flammable liquid already in the container, enough to produce a flammable atmosphere. The operator removed the lid from the container and was about to push the hose into the opening when a flash fire burned his face. Nothing had come out of the hose.

Although the operator was grounded, tests showed that contact between his gloves or outer garment and the container could produce an electric charge on the container large enough to ignite the vapor if it discharged as a spark. A wrench and a loose flange were resting on the container and they could have collected this charge or become charged by induction. The charge from one of them could then have discharged to the metal cage as a spark.

Why did the ignition occur just as the operator was about to insert the hose? (When investigating any fire or explosion, we should always ask why it occurred when it did and not at some other time.) Perhaps the oper-ator, leaning on the container, caused the wrench or the flange to move nearer to one of the bars of the cage, or perhaps the charge passed from the wrench or flange to the operator. Tests showed that such a spark could pass through his clothing.

This incident shows how hard it is to remove all sources of ignition and that the only safe way is to avoid production of a flammable mixture, in this case by inerting the container or perhaps by using a collapsible one so that there is no vapor space.

10.7.2 An Unusual Effect of Static Electricity

A company was filling bags with a powder automatically, using a machine that delivered 50 kg (110 lb) into each bag. Although hand filling showed that this amount of powder would fit into the bags, nevertheless it backed up into the machine and caused it to stall. It had to be stripped down and cleaned before packing could continue.

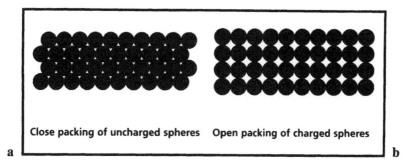

Figure 10-3a,b. (a) Close packing of uncharged spheres. (b) Open packing of charged spheres.

Experiments showed that rapid filling of the powder lowered the bulk density by 10% as charges of static electricity on the particles pushed them farther apart. Instead of being closely packed as in Figure 10-3a, they were more openly packed as shown in Figure 10-3b. As the output of the plant was large, a machine for eliminating the static charge was designed and installed [13].

There is more about static electricity in Sections 2.2.3, 3.2.7, 6.2.5, and 8.1, and in *WWW,* Chapter 15.

10.8 . . . THAT A LITTLE CONTAMINATION CAN HAVE A BIG EFFECT

Some users of X-ray film complained to the manufacturer that there were brown spots on many of the films. The chemist asked to investigate the problem was puzzled. The decisive clue was a comment by another employee that he had not caught any fish in the little local river since a tannery opened 5–6 km (3–4 miles) upstream. The overalls used in the film factory were washed in this water. Could they be the source of the contamination? Tests showed that the water was contaminated with polyphenols and some of it got on the overalls. A speck of fiber carrying only a few thousand polyphenol molecules and blown onto the films by the ventilation system could produce the spots, though they took two months to develop. The laundry passed the water through an ion exchange purifier before using it but the ion exchange resin could not remove polyphenols [14].

A tank truck containing a few inches of used motor oil was brought into a workshop for some welding to be carried out underneath the tank. The welder asked if the contents of the tank were flammable and were told that they were not. When welding started, the tank exploded, causing severe damage but fortunately no injuries. No one realized that used motor oil contains some gasoline, enough to produce a flammable mixture of gasoline vapor and air in the tank, especially on a warm day. This is another example of people not knowing the properties of the material they handled [15].

It was in any case bad practice to weld on equipment containing motor oil. All high-boiling-point oils are flammable though not highly flammable, that is, their flash points are well above ambient temperature. They are not easily ignited but nevertheless there have been many fires and explosions in equipment containing small amounts of such oils because welding has vaporized the oil and then ignited it (see *WWW*, Section 12.4).

Another incident involving used motor oil occurred when welding was carried out between 1.5 and 3 m (5–10 ft) from a small tank containing similar oil. There were several openings in the top of the tank and welding sparks ignited the gasoline vapor. The tank was mounted on wheels so it could easily have been moved if anyone had realized that it contained an explosive mixture [15].

10.9 . . . THAT WE CANNOT GET A TIGHT SEAL BETWEEN THIN BOLTED SHEETS

Section 6.1 of *WWW* describes two incidents that occurred because air leaked into ducts made from thin bolted metal sheets. One occurred in a large blowdown system and led to an explosion that was ignited by the flare. The report recommended that joints between nonmachined surfaces should be welded, that there should be a continuous flow of gas to sweep away any leaks that occurred, and that the oxygen content in blowdown systems should be measured regularly.

The second incident occurred in the same plant 9 mo later because another unit did not carry out the recommendations; perhaps no one told them. A small bolted duct conveyed gland leaks from compressors to a vent stack. Air leaked in to the duct and the mixture of hydrogen, carbon monoxide, and air was ignited by lightning and exploded.

Another incident occurred on a thin metal cabinet containing sparking electrical equipment. As the cabinet was located in a Division 2 area, it was purged with nitrogen. The nitrogen supply became contaminated with a flammable liquid by reverse flow from process equipment at a higher pressure (see Section 9.5); later it failed entirely, air leaked in through the bolted joints, and an explosion occurred, injuring one man [16].

10.10 ... THAT UNFORESEEN SOURCES OF IGNITION ARE OFTEN PRESENT

Many incidents have shown that sources of ignition are liable to turn up even though we have tried to remove all those we can foresee. Elimination of ignition should never be accepted as the basis of safety (unless an occasional explosion is acceptable). Nevertheless explosions still occur because people believe that ignition is impossible.

A vapor-phase oxidation unit consisted of:

1. A vaporizer for the raw material;
2. A mixing chamber where it was mixed with air;
3. A heat exchanger to heat the mixture;
4. A flame arrester; and
5. A tubular reactor (see Figure 10-4).

The reactor operated in the explosive range but below the auto-ignition temperature. The designers realized that hot spots might form in the reactor and ignite the reaction mixture so they strengthened the reactor and provided explosion vents. The flame arrester was installed to prevent the explosion passing back into the heat exchanger. There was no need, they decided, to strengthen or vent the vaporizer, mixer, or heat exchanger as there was no source of ignition in them, or so they thought.

After a 2-yr operation, an explosion demolished the mixer and damaged the heat exchanger. The probable source of ignition was an unlikely one. The vaporizer had to be cleaned from time to time. Various agents had been used including acids, which had attacked the vaporizer and deposited a mixture of metal and organic residues in the mixer. These oxidized and became hot enough to ignite the flammable mixture of reactant vapor and air in the mixer.

When the plant was repaired, the reactant vapor and air were mixed in the reactor, not before. A flammable mixture was then present only in the reactor. This is an inherently safer solution [17]. This could have been

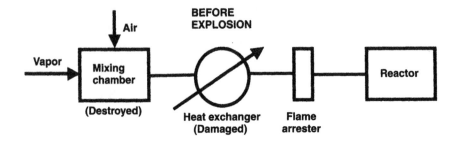

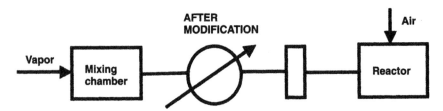

Figure 10-4. Layout of equipment in vapor phase oxidation unit before and after explosion in reactor.

done in the original design if someone had realized that flammable mixtures are easily ignited and that we should therefore avoid the need for them when possible and assume they might explode when their presence is essential. In addition to the necessary repairs, the whole plant was strengthened.

Bond [18] summarizes many fires and explosions caused by unsuspected sources of ignition.

There are other examples of little-known knowledge in *WWW*, Chapter 19, including that ammonia can explode, diesel engines can ignite leaks, mists can explode, and carbon dioxide can ignite a flammable mixture.

10.11 . . . THAT KEEPING THE LETTER OF THE LAW IS NOT ENOUGH

An explosion in the vapor space of a fixed roof storage tank caused complete failure of the wall-floor weld and the whole tank, apart from the floor, rose into the air, leaving the contents behind. They caught fire. One man was killed and 8 others were seriously injured. An adjacent tank also lost its contents. Altogether about 4,000 m³ (1 million gal) of acid was spilled and some of it contaminated a nearby river.

The tank contained sulfuric acid recovered from an alkylation process and contaminated with a small amount of hydrocarbon, enough to produce a flammable mixture in the vapor space. The tank was supposed to be inerted with carbon dioxide but its flow rate was too low to prevent air coming in through various openings in the tank, many of which openings had been caused by corrosion. Welding was taking place above the tank and the probable source of ignition was a spark falling through one of the corrosion holes in the roof or contacting vapor coming out of one of the holes [19].

The following are the main failings that led to the explosion:

• Hot work should not have been allowed so near a tank from which flammable vapor was escaping. There was no periodic or continuous monitoring of the atmosphere.

• The flow of carbon dioxide was too low, either because it was not measured so no one knew what it was or no one had calculated the flow necessary (see Section 9.5). In this case, both were true. The carbon dioxide was supplied by a hose pushed through a hole in the roof. Some of it escaped through corrosion holes and some through the overflow pipe which was shared with tanks vented to the atmosphere.

• The oxygen content of the vapor space was not measured.

• The tank was not provided with a weak seam roof, that is, a wall/roof weld that is weaker than the wall/floor weld so that excessive pressure will cause the wall/roof weld to fail and the liquid will remain in an open cup.

• Thickness measurements and an internal inspection of the tank (and many others) were repeatedly postponed although the company's own inspectors had drawn attention to the need for them and the tank had been emptied several times.

• The dike was big enough to contain the contents of the largest tank within it but it was not designed to prevent a sudden large release from overflowing. Most dikes are the same [20]. Sudden large releases are rare but other cases have occurred and there is a case for increasing dike heights if vulnerable sites such as public highways are near them.

• The company claimed that the various regulations on the storage of chemicals did not apply to the contents of the tank. The managers seem to have believed that following the letter of the law in a hair-splitting way was all that was required. Both the law and the managers were at fault. In contrast, in the UK there is a general requirement to provide a safe plant and system of work and adequate instruction, training, and supervision, so far as is *reasonably practicable* (see the beginning of Chapter 5). The report does not tell us what, if any, training on safety the staff received as students or from their employer.

So many things were below standard at this plant that it is hardly necessary to describe the underlying causes in detail. The senior management of the company seems to have been afflicted by a combination of ignorance and lack of concern. Readers in better-run plants may wonder if there are any lessons for them to learn. However, while it is unusual to find so many faults in one place, each of them has occurred elsewhere on many occasions.

10.12 . . . THE POWER OF COMPRESSED AIR

Contractors were removing water from a pipeline, ≈1 m (3.3 ft) diameter, by pushing a foam pig along the line with compressed air at a gauge pressure of 28 bar (400 psig). The water exit line was rather small, ≈0.3 m (1 ft) diameter so the contractor opened up the end of the pig trap and put a large front end loader in front of it to catch the pig. The force on the pig was so great, nearly 250 tons, that the pig knocked over the loader and traveled another 150 m (500 ft), destroying a wooden platform on the way. Fortunately, no one was standing in the path of the pig at the time.

Many people do not realize the energy in what they call a puff of air or understand the difference between pressure and force. When pressures are measured in pounds per square inch, as they were by the contractor, the full name gets shortened and everyone talks about a pressure of, say, 400 pounds, forgetting or not realizing that this force is exerted on every square inch of the surface. It would be safer to measure pressure in bars or find another name for pounds per square inch.

REFERENCES

1. Mannan, M.S. and H.H. West (1999). Spontaneous combustible substances: a database update. *Proceedings of the Mary Kay O'Connor*

Process Safety Center Annual Symposium, College Station, TX, pp. 267–281.

2. Roberts, R., W.J. Rogers, and M.S. Mannan (2002). Prevention and suppression of metal packing fires. *Proceedings of the Mary Kay O'Connor Process Safety Center Annual Symposium*, College Station, TX, pp. 123–130.

3. Mahnken, G.E. (2000). Watch out for titanium tube-bundle fires. *Chemical Engineering Progress*, **96**(4):47–51.

4. Kelly, M. (2001). *Safety Alert*, Chevron Phillips, Houston, TX, 18 April.

5. Anon. *Safety Alert*, Mary Kay O'Connor Process Safety Center, College Station, TX, 16 April.

6. Urben, P.G., ed. (1999). *Bretherick's Handbook of Reactive Chemical Hazards*, 6th ed., vol. 1, Butterworth-Heinemann, Oxford, UK and Woburn, MA, p. 31.

7. Anon. (2001). Undeclared dangerous goods problems, *Safety Digest — Lessons from Marine Accident Reports*, No. 1/2001, Marine Accident Investigation Branch of the UK Department of Transport, Local Government and the Regions, London, p. 29.

8. Whitefoot, B., quoted by D. Claridge (2000). *Memories: Wilton Castle Club*, ICI Chemicals and Plastics, Middlesbrough, UK, p. 69.

9. Bell, N.A.R. (1971). Loss prevention in the manufacture of nitroglycerine, in *Loss Prevention in the Process Industries*, Symposium Series No. 100, Institution of Chemical Engineers, Rugby, UK, pp. 50–53.

10. Fishwick, T. (2002). Ethylene oxide release sets off alarm system. *Loss Prevention Bulletin*, Oct., 167:21–22.

11. Pratt, T.H. and J.G. Atherton (1999). Electrostatic ignition in everyday operations: Three case histories, *Process Safety Progress*, **18**(4):241–246.

12. Ackroyd, G.P. and S.G. Newton (2002). Flash fire during filling. *Loss Prevention Bulletin*, June, 165:13–14.

13. Pavey, A. (1997/1998). Hidden charges, *Process Safety News*, published by Chilworth Technology, Southampton, UK, No. 4, Autumn/Spring p. 2.

14. Levi, P. (1985). *The Periodic Table*, Michael Joseph, London, pp. 204–208.

15. Ogle, R.A. and R. Carpenter (2001). Lessons learned from fires, flash fires, and explosions involving hot work. *Process Safety Progress*, **20**(2):75–81.

16. Kletz, T.A. (2001). *Learning from Accidents*, 3rd ed., Butterworth-Heinemann, Oxford, UK and Woburn, MA, Chapter 2.

17. Broeckmann, B. (1999). Explosion protection of mixing unit prior to chemical reactor by pressure resistant design, *Proceedings of the Third World Seminar on the Explosion Phenomenon and on the Application of Explosion Protection Techniques in Practice*, Ghent, Belgium, Feb.

18. Bond, J. (1991). *Sources of Ignition*, Butterworth-Heinemann, Oxford, UK and Woburn, MA.

19. Anon. (2000). *Investigation Report: Refinery Incident*, Report No. 2001-05-1-DE, US Chemical Safety and Hazard Investigation Board, Washington, D.C.

20. Thyer, A.M., I.L. Hirst, and S.F. Jagger (2002). Bund overtopping — The consequence of catastrophic tank failure. *Journal of Loss Prevention in the Process Industries*, **15**(5):357–363.

Chapter 11

Control

The main cause of control system failure was inadequate specification.

— *Out of Control* (UK Health and Safety Executive)

11.1 INSTRUMENTS THAT CANNOT DO WHAT WE WANT THEM TO DO

11.1.1 Measuring the Wrong Parameter

The pressure of a water supply was normally high enough for it to be used for firefighting. If the supply pressure fell, a low-pressure alarm sounded and an alternative supply of water was then made available. When someone isolated the water supply in error, the trapped pressure in the line prevented the alarm from operating. The instrumentation could do what it was asked to do — detect a low pressure — but not what its designers wanted it to do, that is, detect that the water supply was unavailable.

As often happens, something else was wrong as well: the valve in the water line should have been locked open but was not. Valves that are locked open for safety reasons should be listed and checked periodically to make sure that they are still locked. They are part of a protective system.

11.1.2 An Alarm That Immediately Reset Itself

A rotameter was designed to measure a gas flow. If the flow stopped or decreased substantially, the float (bobbin) dropped and interrupted a light beam. This triggered a low flow alarm.

The design had limitations. If the flow diminished only slightly, the light beam remained broken and the alarm light stayed on after the alarm bell was silenced. However, if the flow fell substantially or stopped completely, the float dropped farther, the light beam was no longer broken, and the alarm light went out (Figure 11-1). One day when the flow actually failed, the operator canceled the alarm, but with no light to remind him he was distracted by other problems and forgot that the gas flow had stopped. Several hours passed before this was discovered.

Afterwards, the design was changed so that the light beam was broken when the flow was normal but fell on the light sensor if the flow changed (see Figure 11-1b). The alarm light then remained lit as long as the low flow continued. As a bonus, it was also activated by a high flow.

Alternatively, the alarm could have been modified so that once it operated the light stayed on until reset by the operator.

11.1.3 A Trip That Did Not Work Under Abnormal Conditions

Carbon dioxide byproduct from an ammonia plant was sent down a long 1,000-m (3,300-ft) pipeline to another unit. The gas normally contained 2–3% hydrogen. If the hydrogen content rose >8%, contamination by air could produce an explosive mixture. A trip system was therefore installed to shut down the transfer if this figure was reached. The hydrogen level measurement was based on thermal conductivity.

During shutdowns, the ammonia plant was swept out with nitrogen, which contaminated the carbon dioxide. Nitrogen has twice the thermal conductivity of hydrogen so the hydrogen measurements were ignored until the nitrogen had been swept out of the pipeline. You have already guessed what happened: air got into the transfer line during this period and an explosion occurred; 850 m (2,800 ft) of the pipeline was destroyed (Figure 11-2).

The source of the air was never identified. Following an earlier incident, different types of connector were used for compressed air and nitrogen hoses, so compressed air could not have been used by mistake instead of nitrogen for sweeping out the ammonia plant. The source of ignition may have been heat from cutting a bolt.

The report [1] comments, "Looking back it may seem unbelievable. . . . From management and down there had been a will to make safety a priority. During the previous 10 years, considerable money and resources

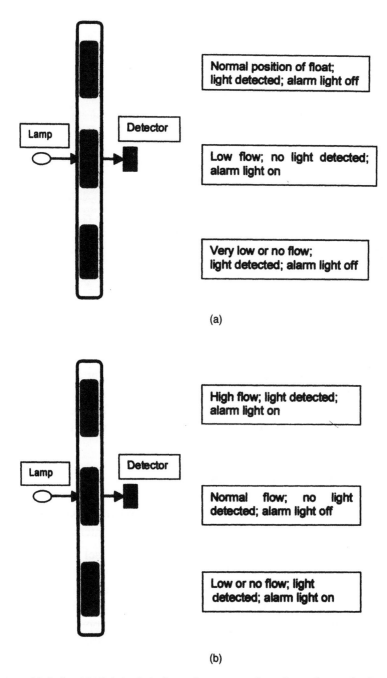

Figure 11-1a,b. (a) Original design of rotameter low flow alarm. (b) Revised design of rotameter low flow alarm.

Figure 11-2. The result of hydrogen and air entering a pipeline. From reference 1. Reprinted with the permission of the American Institute of Chemical Engineers.

had been spent. It was a painful surprise. With hindsight anyone can tell how the explosion could easily have been prevented." Afterwards, the trip system was modified by making use of a carbon dioxide measurement as well as a hydrogen measurement. And the operators were given a better understanding of the problem.

11.1.4 A Sight-Glass with Limited Range

In this and the following two incidents, the laws of physics prevented the equipment from working in the way the designer intended.

A sight-glass 1.2 m (4 ft) long was connected to vessel branches 0.6 m (2 ft) apart as shown in Figure 11-3. It will indicate the correct level only when the liquid in the vessel is between the two branches. If the liquid level is below the lower branch, the liquid in the sight-glass is isolated and its level cannot fall. If the liquid level is above the upper branch, vapor will be trapped in the upper part of the sight-glass. As the level rises, this vapor will be compressed. If there is any noncondensable gas present, the pressure in the sight-glass will rise and the level in the sight-glass will be depressed below the level in the vessel.

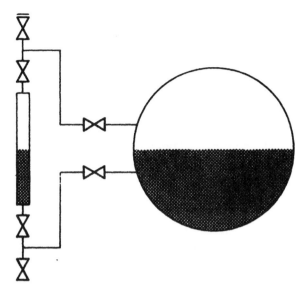

Figure 11-3. A level glass with a limited range. From *Chemical Engineering Progress*, July 1980. Reprinted with the permission of the American Institute of Chemical Engineers.

11.1.5 An Explosion in a Nitric Acid Plant

Ammonia was vaporized, mixed with air, and then passed over a catalyst. The ammonia and air flows were measured and a flow ratio controller was supposed to keep the ammonia concentration below the explosive level (Figure 11-4). The level controller on the vaporizer was out of order and the level of ammonia was being controlled manually. The level got too high and droplets of ammonia were carried forward. All flow measurements are inaccurate when spray is present so the flow ratio controller did not detect the increased flow of ammonia and an explosion occurred. The size of the error in the flow measurement depends on the detailed design; if the spray increases the density of the gas by 50%, the flow of vapor and liquid could be 25% higher than the flowmeter reading.

11.1.6 Vapors and Noncondensable Gases Confused

The following has been discovered more than once during hazard and operability studies. A vessel containing a liquefied gas such as LPG is fitted with a level controller (not shown) and, in addition, a high level trip to isolate the inlet line if the level gets too high (Figure 11-5). The high

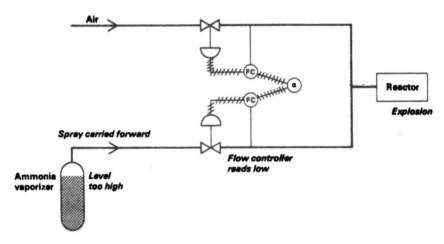

Figure 11-4. An increase in the level in the vaporizer led to an explosion in the nitric acid reactor. From *Chemical Engineering Progress*, July 1980. Reprinted with the permission of the American Institute of Chemical Engineers.

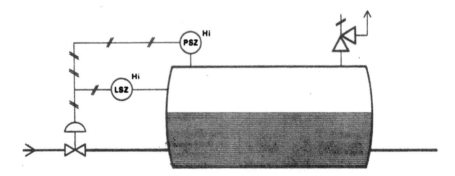

Figure 11-5. The designer of this system seemed unaware of the difference between vapors and noncondensable gases. From *Chemical Engineering Progress*, July 1980. Reprinted with the permission of the Institute of Chemical Engineers.

level trip might fail; the relief valve will then lift and discharge liquid to the atmosphere so a high-pressure trip is installed as well.

If the space above the liquid contains some nitrogen or other noncondensable gas, the system will work. As the level rises, the gas will be compressed and the pressure will rise gradually. However, if there is no noncondensable gas present and the level rises slowly, the system will not work.

The vapor will condense and the pressure will not change until the vessel is completely full of liquid. The pressure will then rise too rapidly for the high-pressure trip to operate and the relief valve will lift.

Condensation takes a finite time. If the level rises rapidly, the vapor may not have time to condense and the system will then work.

The designer of the system probably did not understand the difference between a noncondensable gas, such as air or nitrogen, and a vapor.

11.1.7 Protective Equipment Caused an Explosion

A plastics manufacturing plant included a grinder to eliminate oversize particles. The ground powder was removed by a stream of air. To prevent a dust explosion, there was an explosion suppression system: if a pressure sensor detected a rise in pressure, chlorofluorocarbon (CFC) was released into the grinder and its associated piping to quench the explosion (Figure 11-6).

The system was in use for nearly 20 yr but was never called upon to operate. Then the grinder exploded and the cause was the suppression

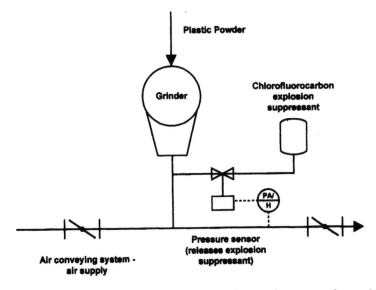

Figure 11-6. Accidental operation of the protective equipment — the explosion suppressant — caused an explosion. From reference 2. Reprinted with the permission of the American Institute of Chemical Engineers.

system. An upset in another part of the unit allowed water to get into the grinding system and form a slurry with the powder. Some of the water or slurry worked its way into the branch leading to the pressure detector. This detector was very sensitive — the pressure exerted by only a few inches of water was sufficient to activate it — so the suppression system operated and the CFC was released. The accumulation of slurry prevented the CFC from flowing easily through the system and the door, weighing over 200 kg (about 500 lb), was blown off the grinder. It hit the wall of the room and bounced back. Fortunately, no one was injured but operators often stood in front of the grinder to inspect its operation through a window in its front. Many people were surprised that the release of CHCs could blow the door off, but it was held by only four bolts and could withstand an internal pressure of only 1–1.4 bar gauge (15–20 psig).

It was certainly a physical explosion and not a chemical one as there was no soot or burnt material and the powder on the floor was still white [2].

The plant was designed before the days when Hazop was widely used. If a Hazop had been carried out on the design, the possibility of water entering the system could have been recognized. (Though Section 8.8 refers to an incident in which none of the Hazop team members recognized that water could get into a unit.)

Today's explosion prevention systems often measure the rate of pressure rise and other materials are used instead of CFCs (because they affect the ozone layer).

It is good practice when designing any equipment to ask, if it were to become overpressurized, which part(s) will give way; as well as to locate the equipment so that people are unlikely to be in the line of fire. We protect equipment from excessive pressure by relief valves or in other ways but no protective system is 100% infallible.

It is, of course, essential to make sure than no one is in the line of fire from equipment that is designed to discharge, that is, relief valves, rupture discs, and pressure vents. The explosion vents on dust handling equipment can produce far longer flames than most people consider possible. An operator was burned when an explosion occurred in a spray dryer and the pressure vents opened [3]. Entrance to the surrounding area was prohibited when the plant was on line but the operator had gone there to look at a noisy pump. Passive protection — in this case, fitting a duct leading to

the outside — is more effective than instructions. If fitting a duct was impossible, then a color photograph of a flame coming out of an explosion vent would have more impact than a written instruction.

In cases like this, it would be a remarkable coincidence if a dust explosion occurred on the one and only occasion that someone defied instructions and went near the vent when the plant was on line. I suspect that the rule had been broken before and that other personnel turned a blind eye.

11.1.8 A Procedure that Cannot Do What We Want It to Do

If tests are being carried out on a vessel, they are often made on the points of a grid. Lines of weakness (such as welds) may then be missed. The grid should be tilted so that the test points are not all above or below each other (Figure 11-7).

Chapter 10 describes mechanical equipment that cannot do what we want it to do.

11.1.9 Preventing similar errors

There is no simple way of preventing the errors described in the foregoing. Hazard and operability studies will help but only if the teams, or at least some of their members, have a good understanding of what is scientifically possible and of the sort of errors that have occurred in the past. The more we discuss our designs with other workers, including those who

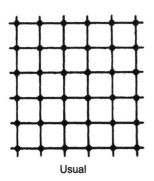

Usual

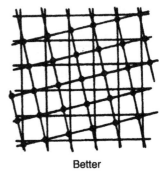

Better

Figure 11-7. If test points are on a grid, it should be tilted so that any lines of weakness are not missed.

will have to operate the equipment, the more likely that someone will spot any weaknesses.

11.2 TOO LITTLE INSTRUMENTATION

A tube rupture and fire in a furnace were the result of too little measurement and control. Two furnaces heated the hydrogen supply to a hydrogenation reactor in four parallel streams (Figure 11-8). The check valve in the combined line beyond the heaters was leaking. When the unit was shut down, liquid flowed backwards from the reactor to the furnaces and settled in a bend in a low point of one stream. This restricted the flow in that stream and the tube got too hot. It did not rupture immediately but expanded more than usual; this caused a small crack and leak elsewhere in the furnace, in the convection section of the furnace. The leak ignited and the flame impinged on another part of the tube, which ruptured. The resulting fire damaged half the tubes. Replacement took six weeks and cost a million dollars but the consequential loss was much greater.

When the furnace was rebuilt, changes were made to the design to reduce the stress in the convection section, more temperature measurements were installed, isolation and control valves were installed in each

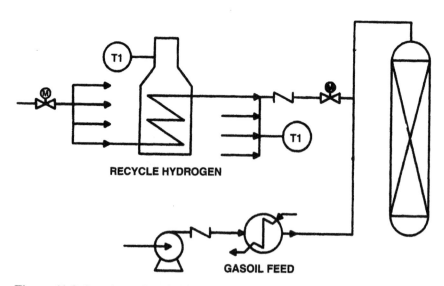

RECYCLE HYDROGEN

GASOIL FEED

Figure 11-8. Restricted flow in one of the parallel streams led to a tube rupture and fire. From reference 4. Reprinted with the permission of Elsevier (UK).

of the four paths, and fire-protected remotely operated emergency isolation valves were fitted so that ruptured tubes could be isolated quickly.

The report [4] does not say whether or not the check valve was inspected regularly. These valves have a reputation for unreliability but few companies schedule them for regular inspection and we can hardly expect equipment with moving parts to operate for the lifetime of a unit without attention.

The plant was constructed in 1978 and seems to have been constructed without many of the features used elsewhere in similar plants at the time, presumably to save money. Remotely operated emergency isolation valves, for example, were widely used from the 1970s [5] onwards.

11.3 DIAGRAMS WERE NOT UP-TO-DATE

Some changes were made to the alarm system on a boiler but no one altered the wiring diagram or made any other record of the changes. Later on, when some other changes were made, the old wiring diagram was used. The operators carried out a few checks to make sure that the low-water-level alarm was working. They confirmed that the warning light came on but they did not check that the burners actually went out. When a low-water level actually occurred during normal operation, the operator was elsewhere. He did not see the warning light, the boiler ran short of water, and it was extensively damaged [6].

There were two major errors in the management system, or rather non-system: The first error was the failure to check all trips and interlocks after a turnaround or modification. At the very least, any equipment that has been worked on should be tested thoroughly. On a furnace, for example, the burners should be lit and a check made that they go out when the water level is lowered. The start-up takes a little longer, but wrecking a furnace causes rather more expense and delay. This is not a new idea. It is a lesson that the company I worked for learned over 40 years ago.

The second error was a common one: a failure to keep line and wiring diagrams up-to-date. Everyone at every company agrees that they ought to do it; many intend to do it, but many more fail to do it. (However, in some countries the law requires it.) Keeping wiring and control cable diagrams up-to-date is particularly important because we can always trace pipelines to see where they go (unless several lines are insulated together), but it is very difficult to trace wires and cables.

11.4 AN AUTOMATIC RESTART FAILS TO RESTART

Following a complete power failure, a rare event expected no more than once every 20–30 years, the emergency supply came on line but one safety-critical pump failed to restart. As a result ≈2.5 tons of chlorine was discharged to the atmosphere through a vent stack. It was discharged at too high a level to cause any harm but it could be smelled over 3 mi away and produced many complaints.

To prevent hunting (i.e., the pump repeatedly switching from one supply to another), the change-over mechanism had to be reset every time it operated. It had been tested several weeks before, by simulating a power failure, confirming that the change-over worked, and then switching the pump back to the normal supply. The tester then forgot to reset the change-over mechanism. This was a foreseeable error but it had not been foreseen. The operators had no way of knowing that the change-over had not been reset. There was an indication in the switch house, but none in the control room. It took the operators 15 min to puzzle out what had happened as none of them really understood how the change-over worked. Do you have any similar equipment at your plant, and do you and your operators know how it works?

Another piece of poor design also misled the operators. A low-flow alarm could have told the operators that the pump had failed. There was such an alarm but it operated the same alarm window as the high-flow alarm. High-flow rates were quite common and not safety critical so the operators did not recognize that the flow was low. They were busy checking that the rest of the plant was okay. Are there any double-duty alarm windows at your plant?

The power failure produced two other learning experiences: Some additional items of equipment needed back-up power supplies; and most operators did not know that they could use their radios when the base station was out-of-action but that they had to use them in a different way.

We can sum up this chapter and others, particularly Chapters 3 and 4, by adapting a computer term and rephrasing the quotation at the beginning of this chapter: "What you should have foreseen is what you get" (WYSHFIWYG).

REFERENCES

1. Pande, J.O. and J. Tonheim (2001). Ammonia plant: explosion of hydrogen in a pipeline for CO_2. *Process Safety Progress*, **20**(1):37–39.

2. Dowell, A.M., D.C. Hendershot, and G.L. Keeports (1999). Explosion caused by explosion suppression system. *Loss Prevention Bulletin*, April, 146:3–4.

3. Anon. (1990). Well-known hazard highlighted by an accident. *Loss Prevention Bulletin*, Oct., 95:6.

4. Kelly, B.D. (1998). Investigation of a hydrogen heater explosion. *Journal of Loss Prevention in the Process Industries*, **11**(4):257–259.

5. Kletz, T.A. (1975). Emergency isolation valves for chemical plants. *Chemical Engineering Progress*, **72**(9):63–71.

6. Donaldson, T. (2001). Control failure incidents. *Loss Prevention Bulletin*, Dec., 162:20–22.

Chapter 12

Leaks

In safety what matters is not who did it, but what was done, and how

— Roger Ford

Section 9.1 of *WWW* quotes figures showing that most leaks occur from pipes or pipefittings (such as valves), often because contractors did not follow either instructions or good work practices when details were left to their discretion. The actions suggested to prevent such leaks included specifying designs in detail and carrying out better inspection during and after construction. This is broadly confirmed by a more recent paper [1] that analyzed 270 leaks on offshore oil platforms. The locations of the leaks were:

Small diameter pipes	18%
Other pipes	43%
Valves	12%
Total pipes & pipe fittings	73%
Vessels	8%
Seals	8%
Pumps and compressors	5%
Hoses	5%
Other equipment	2%

The main immediate causes of the leaks were:

• Corrosion, erosion, and fatigue 32%;

• Wear and tear, such as loss of flexibility in gaskets and valve packing, and friction between moving parts 26%; and

• Poor installation, poor procedures, and failures to follow procedures 39%. Most of these (21%) resulted in an open end; for example, equipment was opened up before the contents were removed.

12.1 LEAKS FROM TANKS

12.1.1 A Leak from a Bad Weld

In 1999, an operator found that a storage tank containing 750 tons of 30% sodium cyanide solution was leaking and that a pool of liquid had formed in the dike. Sixteen tons had leaked but only 4 were recovered as the rest had soaked into the ground. The base of the dike was permeable.

The hazard was primarily an environmental one rather than a safety one as the site is near the River Tees estuary in the UK. Several decades ago, the river was an open sewer but is now home to salmon and seals. Extensive tests showed no harm to wildlife but nevertheless the company was fined for exceeding its discharge authorizations.

The leak was due to the presence of a piece of welding slag, which had been present since the tank was built 22 yr earlier. Water had penetrated between the slag and the weld metal, causing rust to form and creating a leak path.

It is obviously desirable for dikes to have nonpermeable floors but fitting them to existing dikes is very expensive. There are over 150 tanks on this site alone, of various sizes up to $8000\,m^3$ (2 million gal) capacity, and many more throughout the UK. Making the dike floors impermeable would cost $\approx$\$150,000 per dike and would exceed the value of the site [2].

Environmental standards have changed since the site was built and the incident does show the importance, when designing new plants, of asking what changes in safety and environmental standards (and product quality) are likely in the foreseeable future. In some cases, it may be cheaper to meet them now, rather than to pay many times more to modify the plant in the future. In other cases, it may be possible to design a plant so that any equipment needed to meet higher standards later can be added on.

The immediate cause of the loss of material was a poor weld. The underlying cause was failure to provide *options for the future*. By 1977 many people realized that environmental standards were going to be tightened.

12.1.2 A Leak from a Plastic Tank

Hydrochloric acid was stored in a polyester tank that was fitted with a drain valve near the base. Drips from this valve corroded the concrete base on which the tank was sitting, despite a coating of tar, and the loss of support caused mechanical failure of the tank. If that was not enough, the dike (which incidentally was too small) leaked through joints in its walls that had been unsuccessfully sealed with tar. The acid then entered the electrical switch house and contaminated a river. It was neutralized with lime 3 mi downstream [3].

Plastic tanks are often used for corrosive materials. We should remember that they are usually not as strong as steel tanks. There are more incidents involving plastic tanks in *WWW*, Section 5.7.

12.1.3 A Leak from a Lined Tank

Caustic soda was stored in a steel tank lined with rubber. Over the years, rainwater seeped into the gap between the base of the tank and the concrete plinth and caused corrosion, accelerated by the high temperature of the liquid (80 °C, 175 °F). Fortunately, a small leak was noted and the tank was taken out of use and demolished. The outermost foot of the base was badly corroded. The likelihood of corrosion had been noted but nevertheless inspection failed to spot it [3].

12.2 LEAKS FROM LINED PIPES

A leak occurred from a flange on a 1-in pipe, 6 m (20 ft) long, lined with polytetrafluoroethylene (PTFE). This polymer has a much higher coefficient of expansion than carbon steel, 10 × higher averaged over a wide temperature range but up to 75 × higher around 20 °C (70 °F). Unconstrained, a 6 m (20 ft) length will increase by 60 mm (2.4 in) if the temperature rises from 19 °C to 30 °C (66–86 °F) but by only 50 mm (2 in) if it rises from 30 °C to 100 °C (86–212 °F). The pipe was electrically heated and had been temperature-cycled many times.

When the pipe was heated, the liner tried to expand but could not. However, at the higher temperature, the stress was relieved. When the pipe cooled, it tried to contract and pull itself out of the flange. A similar effect occurred in the transverse direction; as the pipe cooled, it pulled itself away from the walls, thus making it easier for lengthwise movement to

take place. In addition, extra trace heating on the flanges, to compensate for the greater heat loss, resulted in thinning of the liner.

After the incident, the company decided to apply less heat to the flanges and to tighten the flange bolts every year. This had been recommended by the manufacturers [4], but had not been done, perhaps because the reason for it was not explained. It is possible that similar effects apply to other plastic linings.

Another leak from a PTFE-lined pipe caused the loss of 4 kg of fluorochlorohydrocarbons, plus smaller amounts of chlorine, hydrochloric acid, and hydrogen fluoride. Small amounts of these gases can diffuse through PTFE and build up behind the lining, cause it to bulge inwards, and restrict the flow. Vent holes were therefore made in the pipe wall to allow the gases to escape. However, combined with the moisture from steam leaks nearby, the gases corroded the pipe beneath the insulation. The lining bubbled-out and failed. The vent holes should not have been covered with insulation. For other examples of gases diffusing through plastic see Section 10.3.

It is well recognized that insulation should be removed from time to time to check for corrosion beneath it (see Section 6.2.1). Places where corrosion is likely should be listed for inspection and some other places picked at random should also be inspected [3].

12.3 A LEAK THROUGH CLOSED VALVES

Most of the nuclear reactors in the UK are cooled by carbon dioxide gas. There is a small loss through leaks and purges and when emptying equipment for maintenance; liquid carbon dioxide is therefore stored on site. It is delivered in tank trucks and pumped through a hose into one of a number of refrigerated tanks. A second hose connects the vapor space in the tank to the vapor space in the tank truck so that they are both at the same pressure. When a load, 15 tons, of carbon dioxide is offloaded $\approx \frac{1}{2}$ ton will flow from the fixed tank back to the tank truck (Figure 12-1).

From the storage tanks, the carbon dioxide is pumped to the reactor cooling system, which operates at a gauge pressure of ≈ 40 bar (580 psi). This gas becomes radioactive. There is also a connection between the reactor and the top of the fixed tanks. It is used to sweep out contaminated gas before maintenance and to remove air afterwards.

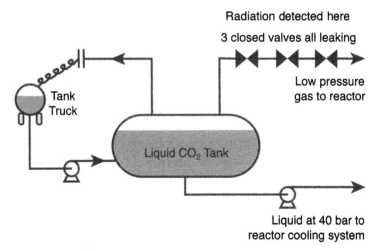

Figure 12.1. Contaminated gas leaked through 3 valves into the fixed tank and from there could have entered the tank truck.

A routine test showed a higher level of radioactivity than usual near this gas line. It was then found that three closed valves in this line were all leaking and some gas was flowing backwards into the storage tanks. It might therefore have gotten into the tank trucks during recent deliveries and contaminated the next load, which might have been delivered to a manufacturer of carbonated drinks. Many tests were carried out, no contamination was found, and calculations showed that even if it had occurred, the dose to the public would have been negligible. Anyone drinking a liter of a contaminated carbonated drink would have received a dose of 1 microsievert (μSv). For comparison, the average background dose in the UK is 2200 μSv per year, many times greater in some areas. Nevertheless, the incident aroused considerable concern among the press, politicians, and public.

According to the official report [5], the root cause of the contamination of the tanks was leaking valves but this was the immediate cause. The root cause was the failure of the designers and operators to be aware of something well known in the chemical and oil industries: a number of valves in series will not provide a positive isolation. For that, a blind, double-block-and-bleed valves, or physical disconnection is necessary. Before entry to a vessel, for example, it is normal practice to isolate all connections by blinds or physical disconnection.

Looking more deeply into the cause, why were the designers and operators unaware that valves are not leak proof? Perhaps it was the insularity of those in the nuclear industry who, like many others, believed that their problems were special and that they could not learn from other industries. Note also that while the immediate cause — leaking valves — is almost certainly correct, the underlying causes are more subjective. They usually are.

After the incident, low-pressure gas for sweeping out the reactor was supplied in a different way, by letdown from the 40 bar supply, and all nuclear power stations were asked to carry out hazard-and-operability studies to see if there were any routes by which their carbon dioxide tanks could be contaminated by radioactive gas.

Note that the incident occurred in a section of the plant devoted to the storage of an inert material. When a plant has serious and obvious hazards, it is a common failing to overlook the hazards in the safer parts of the plant.

The next item, Chapter 1, and Section 13.3 describe other leaks through closed valves.

12.4 A LEAK DUE TO SURGE PRESSURE

Surge pressure, particularly water hammer in steam mains, has caused many failures and large leaks of steam and condensate (see *WWW*, Section 9.1.5). Another incident occurred in a 450-mm (17.7-in) steam pipe operating at a gauge pressure of 13.7 bar (200 psi). The details are complex but the essential features were as follows:

- The steam main went down through a tunnel under a road, rose up on the other side, and was joined by another supply line (see Figure 12-2).

- Following flooding, someone entered the tunnel to inspect the insulation. As the steam trap in the tunnel was blowing, it was isolated before entry was allowed but was not repened afterwards.

- The valve located before the tunnel was reached, as well as the valve on the other supply line, were closed and both were passing. The leak in the first valve filled the dip in the main with condensate and the leak in the other valve maintained a steam bubble in the higher part of the main.

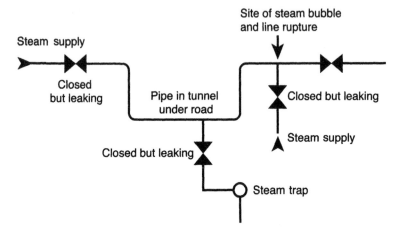

Figure 12.2. Condensate filled the dip in the steam main and overflowed into the horizontal section on the right, condensing a steam bubble. The resultant water hammer ruptured the main.

- Ultimately, the cold condensate completely filled the dip and over-flowed into the horizontal part of the main, causing the steam bubble to condense. The resultant surge pressure ruptured the main at a T-Joint, the weakest part.

Afterward, the company trained over 700 employees on the operation of steam systems. Consultants surveyed the steam system, including 3,000 traps. Over 100 were repaired or replaced and a better system for their inspection and maintenance was set up. However, many other steam mains have also failed due to water hammer. The hazard is well known and has often been described, for example, in the booklet *Hazards of Steam*, first published by Amoco in 1963 and revised in 1984. A similar incident had occurred in the same power station 25 yr earlier. Why did the company not learn from its own and others' experience?

The report [6] does not provide this information, but hopefully the company improved their procedures for reinstating equipment after isolation for entry or maintenance.

12.5 LEAKS FROM SCREWED FITTINGS

During a pressure test at a gauge pressure of $\approx$350 bar (5000 psi), a 20-mm (3/4-in) screwed thermowell was blown out at high speed ($\approx$90 mi/h) and seriously injured a man who was looking for possible leaks. The report [7] does not say whether the failure was due to corrosion, damaged threads, failure to fully engage the threads, or incompatability of the two threads, but all of these have caused other failures of screwed joints (see Section 6.4.2). Many companies do not allow the use of screwed joints except for low-pressure lines handling nonhazardous liquids (hot water is considered hazardous) and for small diameter lines, such as those leading to instruments, and then only after the first isolation valve.

Pressure tests are carried out to confirm that the equipment can withstand the test pressure and, therefore, we should assume that failure is possible and keep everyone out of the way. If we were sure the equipment would not fail, we would not need to test it. Leaks can be detected by testing at the operating pressure.

A screwed nipple and valve blew off of an oil line operating at 350 °C (660 °F). An oil mist 30 m (100 ft) deep covered most of the unit and was sucked into the control room by the ventilation fan. The operators managed to shut down the plant before the oil mist caught fire about 15 min later.

Many people do not realize that mists of flammable liquids can burn or explode at temperatures well below the flashpoint of the vapor. The droplets behave like particles of dust but there is often some vapor present as well and thus these explosions may be more powerful than dust ones.

The nipple that failed was installed during construction to aid pressure testing and was not shown on any drawing. If the operating team had known it was there, they would have replaced it by a welded plug. Afterward, they drew up a list of other weak spots in piping systems to be identified, modified if this was practical and, if not, they inspected the weak points regularly. The following is based on their list [8].

12.6 OTHER WEAK SPOTS IN PIPEWORK

- Vents, drains, and other connections with no obvious function, or which are no longer needed for their design function, should be replaced by welded plugs and as described in Section 6.4.2. If used occasionally, they should be blanked. If used regularly, they should be fitted with double isolations.

- Permanent connections to service lines such as steam, nitrogen, and compressed air should be fitted with check valves and, if the service pressure could be less than the pressure in the process equipment, with double-block-and-bleed valves. A low-pressure alarm should be fitted on the service line (see Section 9.5). If the connection is used only occasionally, and the temperature and pressure are moderate, it may be better to use a hose instead of a permanent connection as long as it is certain that the hose can be vented before it is disconnected. However, hoses generally should be used only for temporary jobs.

- Expansion couplings should be avoided on lines carrying hazardous materials. Expansion loops are less likely to fail.

- Small nipples on main pipes should not be less than 1-in external diameter, although their internal diameter can be less.

- Unusually long runs of small pipe leading to instruments or sample points should be fitted with isolation valves close to the main pipe.

- Inadequately supported small pipes should, of course, be supported.

- Brass valves in process lines are suitable only for low-pressure- and-temperature water lines.

- Unused sample coolers should be removed.

- Equipment (pumps, exchangers, pipes, etc.) no longer needed should be removed.

- Look out for changes in flange ratings in a line. Is the relief setting suitable for the lower flange rating?

- Screwed joints should be avoided, except for small diameter lines containing nonhazardous materials and then only after the first isolation valve (see previous section).

- Cast iron fittings, sometimes found on old units, are brittle and can be broken by impact. Replacement should be considered.

- Unnecessarily large liquid draw-off and sample connections should be replaced or, at least, fitted with restriction plates (or a length of small diameter line, which is less easily removed than a restriction plate).

- Control and electric cables exposed to possible fire damage should be fitted with fire protection as a small fire can cause extensive damage to them and this is expensive to repair.

- Hidden connections under insulation.

- Relief valve tail pipes that discharge to the atmosphere, especially those that could impinge on other equipment if the discharge is ignited by lightning (see Sections 13.1 and 13.2).

- Are control room air inlets located so that they might draw in contaminated air?
- Lines that could cause a serious fire or other incident if they leaked should be scheduled for regular inspection.

Some other leaks are described in Chapter 6.

REFERENCES

1. Health and Safety Executive. (2000). *Offshore Hydrocarbon Release Statistics*, Report No. OTO 200 112, HSE Offshore Division, London, UK; summarized by J.N. Edmondson and D.B. Pratt, Reducing offshore hydrocarbon leaks — A UK regulatory initiative. *Loss Prevention Bulletin*, Dec., 168:4–11.

2. Whitfield, A. (2001). COMAH and the environment — Lessons learned from major accidents 1999–2000. *Hazards XVI — Analysing the Past, Planning the Future*, Symposium Series No. 148, Institution of Chemical Engineers, Rugby, UK, pp. 799–899.

3. Anon. (2002). Corrosion of support structures. *Loss Prevention Bulletin*, Feb., 163:13–14.

4. Schisla, R.M., S.C. Ernst, and P.N. Lodal (2001). Case History: PTFE lined pipe failure. *Proceedings of the 35th Annual AIChE Loss Prevention Symposium.*

5. Anon. (1998). *Contamination of the Carbon Dioxide Supply System at Hunterston B Power Station*, Feb. 1997, HSE Books, Sudbury, UK.

6. Galante, C. and S. Pointer (2002). Catastrophic water hammer in steam dead leg. *Loss Prevention Bulletin*, Oct., 167:16–20.

7. Anon. (2002). *Process Safety Beacon: Faster than a Speeding Bullet*, American Institute of Chemical Engineers, New York, Aug.

8. Anon. (1999). Heavy oil fire. *Loss Prevention Bulletin*, April, 146:20.

Chapter 13

Reactions — Planned and Unplanned

An expert is one who learns through his own experience how painful and deep are the errors one can make even in the most limited field of research.

— Niels Bohr

13.1 DELAYED MIXING

O-chloronitrobenzene reacts with methanol and caustic soda to produce o-nitroanisole:

The reaction is semibatch and operations are normally carried out as follows:

1. The first two reactants are placed in the reactor and mixed.

2. The stirrer is then switched off and the liquid level checked by opening the manway cover.

3. The cover is replaced and the stirrer is switched on.

4. The temperature is raised to 80 °C (175 °F) by passing hot water through the reactor jacket and the pressure is raised to 9 bar gauge (130 psig) with nitrogen.

5. A solution of caustic soda in methanol is then added gradually and the temperature kept at 80 °C (175 °F) by adjusting the flow rate of cold water through the cooling jacket.

One day, after replacing the manway cover (step 3), the operator forgot to switch the stirrer back on. There was no mixing and the caustic soda plus methanol (added in step 5) formed a separate layer. No reaction occurred and the operator had to apply heat instead of cooling to the jacket to maintain the temperature at 80 °C (175 °F). When the operator realized that the stirrer was not running, he switched it on. A very rapid reaction occurred, the temperature rose to at least 160 °C (320 °F) and the pressure to 16 bar gauge (230 psig). Most of the contents of the reactor, 10 tons of liquid, were discharged through the relief valve and a yellow deposit was distributed over 300,000 m^2 (75 acres) of a built-up area, a suburb of Frankfurt, Germany.

At one time, an accident like this would have been blamed on human errors — forgetting to switch the stirrer on after replacing the cover and then switching it on too late. But both errors are easy to make, particularly the second one: when we find we have not done something we should have done, our natural tendency is to do it at once. "Better late than never" is a common saying. However, in this case and many others it proved disastrous (see also Sections 8.5, 13.2, and 13.4). One of the biggest causes of runaways in batch and semibatch reactions is failing to start the stirrer or circulation pump and then starting it late so that large amounts of reactants are suddenly mixed (see *WWW*, Sections 3.2.8 and 22.2). So how can we prevent them?

There were several weaknesses in the design:

- It should not be necessary to open the reactor to check the level. A level indicator or load cell would have removed the need to switch off the stirrer.

- An agitation detector could have stopped addition of the methanol/caustic soda mixture if the stirrer was not running. (This would be better than checking the voltage applied to the stirrer motor as the motor or its coupling can fail.)

- A catchpot after the relief valve would have collected the discharge and prevented it from spreading over the surroundings. This incident occurred in 1993. Nearly twenty years earlier a discharge at Seveso in Italy had taught us the same lesson (see *WWW*, Section 21.2.5). On that occasion, a large area around the Italian plant had been sprayed with dioxin and $4 \, \text{km}^2$ (1,000 acres) made unusable. At the plant involved in the more recent incident, a catchpot had not been fitted because the normal products of the reaction were not all that hazardous. (The caustic soda was well diluted.) However, the designers had overlooked the fact that once a runaway starts, other reactions occur and different and more harmful products are formed.

These design errors could have been avoided if the company had been aware of a similar runaway that occurred in Japan over 20 years earlier and injured 9 people. The published report on it was very brief but it should have been sufficient to alert the German company to the hazard. An underlying cause of the accident was thus a failure to learn from the past, from both Seveso and Japan.

Following the discharge, the adverse reports in the media caused the company to withdraw from the chemical industry. The German authorities imposed further regulations on the industry, affecting all companies [1]. To the public, the chemical industry is a single unified entity and an accident in one company affects others. For this reason, we should share information on the cases of our accidents and the actions we have taken to prevent them from happening again (see Section 16.8).

Another frequent cause of uncontrolled rises in temperature is dissolving caustic soda in water. If the solid caustic soda is added too quickly and/or with insufficient circulation, some of the solid accumulates at the bottom of the mixing vessel and then slowly dissolves, forming a strong solution. Any sudden mixing causes rapid production of heat, local boiling, and further mixing. It is not necessary to switch on a stirrer to start the mixing process; mechanical shock or vibration may be sufficient [2].

A bucket containing 25% sodium hydroxide solution was used to collect bromine that was dripping from a leak. Unreacted bromine formed a separate layer at the bottom. When the bucket was moved the two layers mixed and there was a violent eruption [3].

13.2 WAITING UNTIL AFTER THE FOURTH ACCIDENT

A mixture of phenol, formaldehyde, and sulfuric acid — the raw materials for the manufacture of PF resin — was discharged onto a roadway four times before the company decided to install a catchpot after the reactor rupture disc.

The first discharge occurred because the operator forgot to add the catalyst — sulfuric acid — at the beginning and then added a larger amount later when a second addition of catalyst was normally made. This was another example of the incorrect belief that it is better to carry out an action late than not carry it out at all (see Sections 8.5, 13.1, and 13.4).

The second discharge occurred because the formaldehyde failed to react, for an unknown reason. When the second addition of catalyst was made, the large excess reacted vigorously.

The third and fourth incidents had similar causes. Part of the heat of reaction was removed by a cooling jacket and part by condensing the vapor given off during reaction. The latter was ineffective, as there was a partial choke in the vapor line where it entered the condenser.

The company did not ignore the first three incidents. They changed the operating procedures. After the fourth incident, they decided that was not enough and they made a change in the design: they installed a catchpot [4].

When a hazard is recognized, by experience or in any other way, the most effective action we can take is to remove it. If that is not possible, we can add on equipment to control it. However, relying on procedures should be our last resort. Moreover, in some companies the default action is to think of procedures first — perhaps because they are cheaper and quicker to install and do not require any design effort — but they are less effective. There are other examples of this in Chapter 5.

13.3 LOWER TEMPERATURE MAY NOT MEAN LESS RISK

In the incident described in Section 13.1, the raw materials did not react because there was no mixing. Raw materials can also fail to react for another reason: because they are too cold.

An aromatic amine was reacted with sulfuric acid and nitrosyl sulfuric acid (NSA) to form a compound, which was then decomposed to form a phenol. A hundred batches had been made every year for several years without incident until a runaway reaction occurred. It produced a large amount of gas, which overpressured and ruptured the 2,270-l (600 gal) reactor. It was driven through the concrete floor while its lid traveled 150 m (500 ft). Fortunately, no one was injured.

The reaction was semibatch. The amine and sulfuric acid were mixed in the reactor and then the NSA was added gradually. When reaction was complete, the mixture was moved to another vessel where it was decomposed to the final product.

The heating and cooling of the reactor were temporarily done manually. Probably for this reason temperature control was more erratic than usual and at times the reactor was too cold for reaction to occur. About 30% of the NSA failed to react. The temperature then rose above the normal, probably because the valve supplying steam to the reactor was leaking or had not been fully closed. There are so many probables because data recording was rudimentary.

The replacement plant designed and built after the explosion included computer control, data logging, trips and interlocks, and a quench tank filled with water into which the contents of the reactor could be dumped if they got too hot [5].

Note that several hundred batches were made without incident before the runaway occurred. A blind man can walk along the edge of a cliff for a long time before he deviates from the correct path far enough to fall over the edge. Section 12.3 describes another accident due to leaking valves.

13.4 FORGETTING TO ADD A REACTANT

The reaction between phenol and formaldehyde to form phenol formaldehyde resins has produced many runaway reactions, most of them the result of the same omission. There are two reaction steps. In the first step, phenol is reacted with formaldehyde in a stirred semibatch reactor, which can be heated or cooled. The phenol is charged to the reactor, a small amount of caustic soda is added as a catalyst, and the formaldehyde is then added gradually. In the second step, much more caustic soda is added gradually, this time as a reactant.

A common error is to forget to add the caustic soda in the first stage. The phenol and formaldehyde do not react but the operators do not realize this as the automatic temperature control keeps the reactor at the correct temperature. When the addition of caustic soda starts, there is a violent runaway reaction, which may burst the reactor.

To prevent this occurring, the cooling load should be measured during the first step and the addition of formaldehyde stopped if it is too low. It should also, of course, be stopped if there is no agitation (see Section 13.1) [1]. Another possibility is to carry out the two stages in two different reactors. In the first reactor, it would be possible to add only small amounts of caustic soda.

For the accidents described so far, the solutions suggested are *add-ons*: more protective devices are proposed, devices that may be neglected or switched off. Could the reactions take place in continuous reactors made from long thin tubes? Has anyone looked for alternative and safer chemical processes? Chemists have been slower than chemical engineers to adopt inherently safer designs.

An experienced process designer writes [6]:

I have been involved in the process design of many chemical processes. Quite often, I have been given a technology transfer package and told to design a suitable plant. When I informed my management that the process was hazardous . . . and that it should be modified to be safer, I was then told that it was too late and that too much time and money had already been expended, and that I should use as many safety measures and as much equipment as necessary to make the process safer.

Based on my often frustrating experiences with a *fait accompli* process, I feel strongly that the concepts of *inherently safer design* should be taught at the undergraduate chemical engineering and chemistry curricula level. It may be even more important for chemists to become aware of this technique, as they are the ones who conceptualize and develop chemical processes. If they were aware of the technique, they might come up with inherently safer processes from the start. . . . This would result in lower initial plant costs and fewer accidents, which then would save replacing of equipment and prevent both business interruption and lawsuits.

13.5 INADEQUATE TESTS

Runaway reactions have occurred because the tests carried out were in some way inadequate [7]. One occurred because the sample was very small — a few milligrams — and was not representative of the reaction mixture. The damage was catastrophic. Afterward, tests were carried out with 5–10-gm samples and incidents occurred. It is, of course, easy to get a representative sample of a pure compound but not of a mixture. I remember reading this in my university textbook on chemical analysis.

Another incident occurred because the test measured the heat release but did not measure the amount of gas produced. The process worked in the laboratory and in a pilot plant but when transferred to full-scale operation the vent was too small. The reactor cover lifted and the escaping gas ignited.

A third incident was somewhat similar. The process required two reactants to be mixed and then heated to 85 °C (185 °F). A third reactant was then added slowly. Tests showed the mixture of the two reactants was stable and so they were premixed in drums ready for charging. Two hours later the drums started to rupture. The tests had failed to show that gas was slowly produced even at room temperature. Tests should measure the pressure produced as well as the heat produced.

The onset temperature for a runaway is not a fundamental property like the boiling point and can vary by as much as 50 °C (90 °F), in some cases as much as 100 °C (180 °F), depending on the method used.

The need for thorough testing is shown by a fourth incident. It occurred in a process for the nitration of an aromatic compound by nitric acid in acetic acid solution using sodium nitrite as catalyst. The solution was dilute and tests showed only a moderate rise in temperature. However, in further tests, the reaction mixture was allowed to stand at 70 °C (160 °F) to make sure the reaction was complete. Heat production continued; the temperature reached 180 °C (355 °F) and the gauge pressure reached 25 bar (360 psi). A literature review showed that the excess nitric acid was reacting with the solvent to produce acetyl nitrite, which decomposes at 70 °C (160 °F).

The plant was advised to install extra cooling capacity, triggered by a high-temperature measurement, but in the longer term to look for a less reactive solvent. The experience does show that simple screening of the raw materials and the reaction is not enough [8]. Reaction mixtures are always liable to be left, for any number of reasons, when reaction is complete — or sometimes when it is only partly complete. The most notorious example of the latter is the explosion at Seveso, Italy in 1976, where a partially reacted mixture was left to stand over a weekend [9].

13.6 A HEATING MEDIUM WAS TOO HOT

A product was vaporized and condensed to improve its purity. Vaporization took place in a small jacketed vessel (2.3 m³, 600 gal). It was kept under vacuum and the contents were heated to 140 °C (285 °F), the boiling point under the vacuum, by oil at 170 °C (340 °F). The process had to be shut down for planned maintenance of the steam supply. The vacuum was broken and cooling applied to the jacket but nevertheless the temperature reached 160 °C (320 °F), the temperature at which the product started to decompose, and then rose rapidly. The glass exit pipe broke and the escaping liquid caught fire. A high-temperature trip should have automatically switched the jacket from heating to cooling before the temperature reached 160 °C, but the sensing device was fouled with tar and read low.

The order of events is not entirely clear but it seems that breaking the vacuum stopped the evaporative cooling and that this took place before the cooling agent replaced the hot oil in the jacket (or possibly before the cooling had time to become effective). One wonders if the operators understood that evaporation provided a cooling effect, which was lost when the vacuum was broken.

The report does not say how often the high-temperature alarm was tested but after the incident multiple temperature probes were fitted to the vessel. The report [10] does not mention the major weakness in design: using a heating medium hotter than the temperature at which decomposition started. This is inherently unsafe and the incident shows the inherent weakness of relying on an active protective system (which could and did fail) instead of an inherently safer design. Of course, a cooler heating medium would have needed a larger heating area. A thin film evaporator might have been the best way of achieving this.

13.7 AN UNSTABLE SUBSTANCE LEFT
STANDING FOR TOO LONG

As described in Section 13.5, runaway reactions have occurred because mixtures of raw materials, intermediates, or products have been left standing for too long. This can also occur with single substances. A peroxide was moved from a weigh tank through a transfer pipe to a reactor and the pipe left empty. One day the pipe was left full of liquid while a leak was repaired. The repair took longer than expected and the heat from the reactor slowly warmed the liquid in the pipe until it decomposed and ruptured the pipe. The report [11] says, "Luckily, there were no injuries, just a lot of surprised people." I expect they knew the decomposition temperature of the peroxide. What probably surprised them was that the peroxide could get hot enough by conduction along the pipe from the reactor. We all know that metals are good conductors of heat but most of us have no instinctive grasp of the rate at which heat can flow (or of the rate at which vessels will cool; see Sections 3.1.2, 4 and 5, and 8.7).

REFERENCES

1. Gustin, J.L. (2002). How the study of accident case histories can prevent runaway reaction accidents from recurring. *Process Safety and Environmental Protection*, **80**(B1):16–24.

2. Cox, J. (1998). Caustic layering — the forgotten hazard. *The Chemical Engineer*, 8 Oct., 667:25–28.

3. Anon. (1999). Case Histories of Accidents in the Chemical Industry, No. 1636, quoted by Urben, P.G. (ed.), *Bretherick's Handbook of Reactive Chemical Hazards*, 6th ed., Butterworth-Heinemann, Oxford, UK and Woburn, MA, p.108.

4. Gillard, T. (1998). Loss of reactor contents to atmosphere. *Loss Prevention Bulletin*, Oct., 143:21–22.

5. Partington, S. and S.P. Waldram (2002). Runaway reaction during production of an azo dye intermediate. *Process Safety and Environmental Protection*, **80**(B1):33–39.

6. Grossel, S.S. (2003). Safety issues. *Chemical & Engineering News*, March 17, p. 4.

7. Singh, J. and C. Sims (1999). Reactive chemical screening — A widespread weak link. *Proceedings of the Mary Kay O'Connor Process Safety Center Annual Symposium*, College Station, TX.

8. Roe, S. (1997/1998). Are your process materials fully compatible? *Process Safety News* (published by Chilworth Technology, Southampton, UK), No. 4, Autumn/Spring, p. 2.

9. Kletz, T.A. (2001). *Learning from Accidents*, 3rd ed., Butterworth-Heinemann, Oxford UK and Woburn, MA, Chapter 9.

10. Bickerton, J. (2002). Fire in a vacuum still. *Loss Prevention Bulletin*, Aug., 166:12–13.

11. Anon. (2003). *Process Safety Beacon: Reactive Chemistry: Not Always When or Where You Want*, American Institute of Chemical Engineers, New York, March.

Chapter 14

Both Design and Operations Could Have Been Better

You can't solve problems with the same level of knowledge that created them.

— Albert Einstein

Many of the incidents in this book could have been prevented by better design, and many by better operations. Good operations can sometimes compensate for poor design and vice versa, but that is not something on which we should rely.

14.1 WATER IN RELIEF VALVE TAILPIPES

My first example is a very simple one. Sections 13.1 and 13.2 describe several incidents in which relief valves discharged process material directly into the atmosphere instead of into a catchpot or other closed system such as a flare stack or scrubber. These were hardly unforeseen incidents. Relief valves are designed to lift, so we should not be surprised when they do and we should design accordingly. Relief valves on steam systems can, of course, safely discharge into the atmosphere but they are not without hazards. To prevent steam condensing in the tailpipe and filling it with water, a small hole is usually drilled in the tailpipe as a water drain. However, these holes often get blocked with rust and other debris, and then slight leaking of the relief valve and/or rain causes water to accumulate. This raises the pressure at which the relief valve lifts and when it does a slug of water is blown out. If the relief valve is leaking slightly, this water will be hot.

These drain points are protective systems, and like all protective systems they should be inspected regularly, in this case by rodding to make sure they are clear [1]. I suggest they should be at least 1 in diameter.

Some companies fit drain pipes to the drain holes but this can make matters worse. A long narrow tube can choke more readily than a hole. If drain lines are fitted, they should be short, straight, at least 1 in internal diameter and designed so that they can be checked to make sure they are clear [2].

14.2 A JOURNEY IN A TIME MACHINE

This accident occurred in 1998 but shows such a lack of good practice in both design and operations that I looked at the cover of the report [3] to make sure that it really happened then and not in 1898.

A new unit, alongside an existing one, separated gases from crude oil in three stages. The first-stage separator was designed to withstand a pressure of 95 bar gauge (1,400 psig), and the second a pressure of 35 bar gauge (500 psig). However, operating pressures were lower, 68 bar gauge (990 psig) and 15.5 bar gauge (225 psig). Both separators were fitted with relief valves. The third-stage separator was designed to operate at atmospheric pressure and had neither relief valve nor vent, though a gas exit line and isolation valve were fitted to the top of the vessel. Bypass lines with valves were fitted around all three separators (Figure 14-1). The crude oil was piped from a well about 3 km (2 miles) away.

The new equipment, apart from this pipeline, was freed from air using crude oil from a nearby well. The next job was to sweep the 3-km pipeline free from air using oil from the distant well. The valves were set as shown in Figure 14-1. Note that the two valves in the third-stage bypass should have been open but were shut. No one knew when they were shut or who shut them. The pressure of the crude oil supply pump, designed to pump the oil through a 3-km pipeline, ruptured the separator, killing four people and causing considerable damage to other equipment.

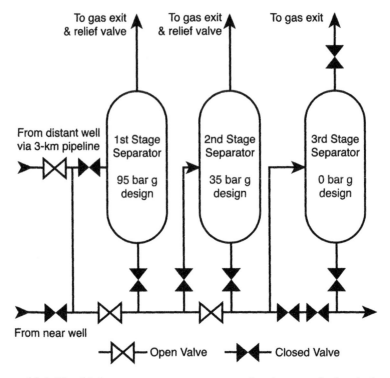

Figure 14-1. The third separator was overpressured and ruptured when its bypass valves were closed in error.

14.2.1 Design Errors

- The major error was the lack of a relief valve on the third separator.

- The vessel could have been protected by an isolation valve in the inlet line but this alone would not be considered adequate. It would be an adjunct to a relief valve, not a substitute for it.

- A hazard and operability study or review of the relief system would have disclosed these design errors.

- There were no drawings! (Compare the explosion at Flixborough in 1974 where the only drawing for the modification that failed was a full-size sketch in chalk on the workshop floor.) (See *WWW*, Section 2.4 and Reference [4].)

We do not know if the design engineers, who were company employees, were unaware of the codes for vessel design or simply decided (or were told) to do a cheap job. However, the savings in cost were miniscule.

14.2.2 Operating Errors

- There were no written instructions for start-up or normal operation. The company policy was to rely on on-the-job training without checking that the messages had been received and understood. Unfortunately, each time procedures or knowledge are passed on, they are liable to be degraded. In verbal communication — and sometimes in written communication — the message sent and the message received are not always the same.

- The valves should have been checked by a responsible person, using a checklist, before purging started.

- The company held monthly safety meetings for all employees but the report does not say whether these covered process safety or just "hard hat" safety.

- Employees frequently moved between different plants belonging to the same company but received no training on the different designs and procedures.

- Sweeping out pipelines and other equipment with flammable liquids is not good practice. It is safer to first inert the lines with nitrogen. Crude oil has a high conductivity and static electricity is therefore not a hazard but other sources of ignition are possible, though unlikely. The American Oil Company booklet, *Hazards of Air* [5], first published in 1958, includes pictures of an underground crude oil pipeline in which detonations occurred along a length of 50 km (30 mi). In this case, compressed air was being used to sweep out crude oil. The pictures show soil blown out and projecting pieces of pipe at intervals of some tens of meters.

We do not know if the company managers were amateurs, and thus unaware of the need for good design and operating methods, or trying to do everything on the cheap. It may have been a mixture of both.

14.3 CHOKES IN FLARE STACKS

Many explosions have occurred in flare stacks because a normally continuous flow of gas failed or fell to a very low level and air diffused down the stack, forming an explosive mixture (see *WWW*, Chapter 6). To reduce such diffusion, many companies have installed molecular seals: the gas leaving the stack and any air diffusing down follow a labyrinthine path, as shown in Figure 14-2. However, these seals have disadvantages. Carbon from incompletely burnt gas can fall into the bottom of the seal and block the flow. Ideally, material discharged from a relief valve should follow a simple and straightforward path without any equipment that might obstruct the flow in the way. For this reason, some companies have had second thoughts and removed the top of the molecular seal, called the "top hat," as shown by the dotted line in Figure 14-2. This effectively neutralizes the seal but leaves the rest of it in position. On one plant, a steam nozzle was added inside the stack to cool the tip and to reduce smoke formation. Some of the steam condensed and flowed down the drain line into the knock-out drum. In the stack shown in Figure 14-2, the drain line was insulated with the steam line to prevent freezing.

Unfortunately, the drain line was only 1 in diameter. It became partially blocked with carbon and a pool of water formed in the bottom of the molecular seal. It filled the outer rings of the seal and overflowed into the inner pipe. Some of it froze when unusually large quantities of cold gas had to be flared during cold weather, partially choking the inner pipe. Finally, there was a rumbling noise and bits of ice and a stream of water were blown out of the stack. The flame was extinguished. Similar incidents have occurred on other plants. On one plant, a narrow stack was completely blocked by ice when steam was injected inside it (see *WWW*, Sections 2.5 and 6.2). It is doubtful if a 107-cm (42-in) diameter stack could be completely blocked in this way. Nevertheless the report [6] recommended installation of a much bigger drain line. Removal of the molecular seal — a major job — would not have prevented icing but all the water formed would have fallen to the bottom of the stack, another example of the unforeseen effects of change.

The major weakness in design was the 1-in drain line, which was far too small for such an important duty. The operating error was not to consider critically the possible effects of fitting a steam nozzle in the stack.

14.4 OTHER EXPLOSIONS IN FLARE STACKS

For an explosion, we need fuel, air (or oxygen), and a source of ignition. In a flare stack, the fuel is almost always there, as the purpose of the flare stack is to burn it; thus as the flare is normally there, all we need is

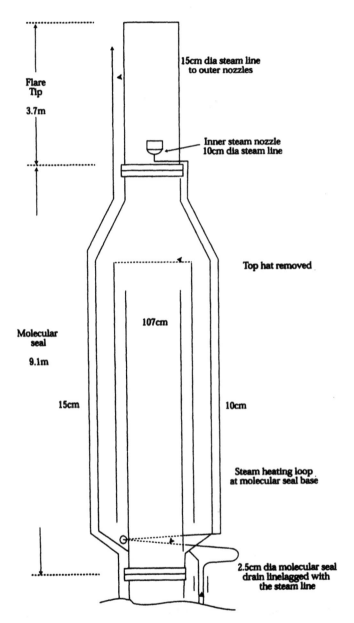

Figure 14-2. The top of a flare stack fitted with a "top hat." From reference 6. Reprinted with the permission of the Institution of Chemical Engineers.

the air. Air can leak into the flare lines from the equipment that feeds the lines or through leaks in the lines leading to the stack, or can diffuse down the stack if the upflow stops or becomes very low (see *WWW*, Chapter 6). Another possible cause is development of a vacuum in the flare system so that air is actually sucked into the stack. This is unlikely but the following are accounts of two explosions that were caused this way [6].

In the first incident, the pressure control valves on two compressor suction drums should have been set to prevent the pressure falling below 2 bar gauge (30 psig) but were set in error at zero. One of them was probably set slightly below zero. This allowed a slight vacuum to form and it sucked air down the flare stack. Calculations showed that if the control valve was open for only a few minutes, the stack would be completely filled with air and thus it would fill with a flammable mixture of vapor and air in less time. In addition, the nitrogen purge, intended to prevent back flow of air down the stack, was only a third of its normal rate. The resulting explosion deformed the base and blew off parts of the tip. They landed 45 m (150 ft) away.

The second explosion occurred on a plant in which some equipment operated under vacuum. There were rupture discs below the relief valves on a section of the plant that was under a slight pressure but they had failed and the relief valves were leaking. This leak increased the vacuum elsewhere in the plant and pulled air into the stack, despite the presence of a molecular seal and a flame arrester in the stack. The initial explosion was followed by two others while the plant was shutting down. The stack was ruptured in three places.

The report [6] suggests that there may have been a split in the top hat of the molecular seal and that the flame arrester should have been nearer the top of the stack. Flame arresters are more efficient when they are near the end of a pipe but they are liable to become dirty and produce a pressure drop. They should not be installed in flare stacks that, as already stated, should provide an uninterrupted path to the atmosphere.

14.5 DESIGN POOR, PROTECTION NEGLECTED

Two large pumps handled a slurry of catalyst and a hydrocarbon similar to gasoline in its physical properties. Two pumps, originally one working, one spare, were operated continuously to increase throughput. However, through put was reduced when one of the pumps had to be maintained.

The pump seals were flushed with clean hydrocarbon from one of the plant vessels.

As the result of an upset, the level in this vessel was lost. For the plant as whole, this was no more than a minor disturbance but the loss of the flush caused the pump glands to leak. The leak was small so it was decided to repair the seals one at a time. The first repair took eight hours, rather longer than expected. Meanwhile the leak on the other seal gradually worsened and was dispersed with a steam lance. Finally, just as the repair of the first pump was completed, the shift foreman, who had just come on duty, decided that the leak was so bad that the plant should be shut down at once. Unfortunately, the valve actuators on the leaking pump's isolation valves had been removed for repair, there was no easy way of operating the valves manually, and there were no alternative valves that could be closed instead. As a result, the leak could not be stopped immediately but continued until the gauge pressure had fallen to zero. Fortunately, the leaking liquid did not ignite.

14.5.1 What Went Wrong?

- It was a serious design error to depend on an unreliable source for the flushing of the pump seals. There should have been either an alternative supply or a more reliable one. If a Hazop had been carried out, then this weakness would have been discovered, provided that the team had the adequate experience and they studied the minor lines as well as the main ones. In some plants, minor lines such as flush lines to pump seals, drain lines, and sample lines are not given a line number and are then overlooked during the Hazop (see Section 14.8).

- There were flowmeters with alarms on the flush lines but they were out of order. If they had been in working order, the loss of flow might have been noticed sooner and damage would have been less.

- It was a major operating error to remove the valve actuators, even for a short time, with the pump on line and without providing a means of manual isolation. This was not forbidden in the plant operating or safety instructions. It was not forbidden because the authors never considered that anyone would want to carry out such a foolish act and probably never realized that the valves could not be hand operated. This incident, similar to a number of others, shows the limitations of instructions. They cannot be a substitute for an understanding of the basic principles of safety, scientific knowledge, or plant operation. We cannot forbid folly in all its possible forms (see Sections 2.5.2, 7.4, 7.5, and 8.12). The report does not say at what level the decision to remove the actuators was taken.

• This incident occurred in a company with a commitment to safety and a good record but at a time when the record had worsened as result of rapid expansion and the construction of many large new plants, larger than those previously operated and operating at higher temperatures and pressures. The need to pay more attention to process safety had been recognized five years earlier and much had been done, but it takes time to change long-established working practices.

14.6 SEVERAL POOR SYSTEMS DO NOT MAKE A GOOD SYSTEM

This incident was the result of too much complexity in design and operations. Unfortunately, this makes the following account rather complex. Please persevere; it contains some messages of wider applicability than you might think at first glance.

Equipment contaminated with radioactivity was cleaned in a shielded room known as a control cell (Figure 14-3). The contaminated equipment

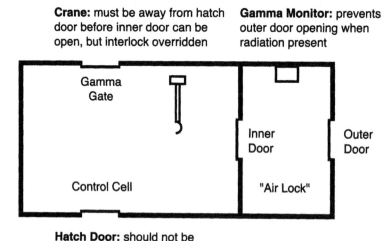

Figure 14-3. An overly complex system of interlocks to prevent both doors being open at the same time, unless maintenance was in progress.

was brought into the cell through a hole in the floor (the hatch door) and removed after cleaning through a hole in the ceiling (the gamma gate). Cleaning and other operations were carried out by remote control. No one was allowed in the cell when radioactive equipment was present but people could enter the cell at other times to maintain the cleaning equipment. They entered through double doors that acted as an air lock. A key exchange system — only one key could be withdrawn at a time — ensured that only one door could be open at a time.

The cleaning equipment needed much more maintenance than expected and the company decided that both doors should be open during maintenance so that escape, if found to be necessary, could be rapid. To permit this, an override was fitted to the key exchange system and at about the same time extra layers of protection were added. In addition to the key exchange system, now fitted with an override, there were an additional six layers of protection:

(a) If the inner door was open and radioactive equipment was present, a gamma ray monitor in the space between the doors prevented anyone from opening the outer door. This interlock was hardwired.

(b) A software interlock prevented the opening of the inner door unless the crane was in such a position that it could not open the hatch door.

(c) Another software interlock prevented the opening of the hatch door unless the inner door was closed. Both software interlocks formed part of the programmable electronic system (PES) that operated the other controls. However, an unknown software error made this interlock ineffective. Testing had not detected the fault.

(d) A hardwired interlock prevented the opening of the hatch door when the override was in operation.

(e) A checklist had to be completed before the doors were opened.

(f) A permit-to-work had to be completed before anyone entered the cell.

Despite all these precautions, both doors were found open when radioactive equipment was in the cell. Fortunately, no one was in the path of the open doors. If they had been, they could have received a serious dose of radiation.

While preparing the plant for maintenance, the foreman found that the inner door would not open. The foreman did not realize that the crane was in the wrong position and that interlock (b) was keeping the door closed. After the maintenance department had spent most of the day trying to determine why the door would not open, the shift manager decided to override interlock (b) by making the PES "think" that the crane was in the right position. He noted what he had done in his log.

Several days passed before the maintenance in the cell was complete. A different team was then on shift and they had not read all the old logs. By remote control, they opened the hatch door although both cell doors were open and moved active equipment into the cell. Interlock (c) would have prevented this happening if it had not been faulty; interlock (b) would have prevented this happening if it had not been overridden. Interlock (a) was ineffective because there was no radiation present when the outer door was opened. The interlock was not designed to sound an alarm if radiation subsequently appeared. Fortunately, the crane operator noticed on his TV screen that the cell doors were open and he was able to close the inner door.

14.6.1 What Went Wrong?

When the inner door would not open, the foreman assumed it was faulty. He should have checked the state of all the interlocks. Afterward, a number of interlocks elsewhere on the plant were found to be overridden. Assuming instruments to be faulty is a common failing. For example, when a high-level alarm on a tank sounds, many operators have said, "It can't possibly be full," and sent for the instrument technician. By the time he arrived, the tank was overflowing.

When the maintenance was complete and the plant ready to be brought back into use, no one visited the scene to check that the outer door, at least, was shut. Again, many operators do not realize that a walk around the plant may reveal something that instruments cannot. In contrast, I read some years ago that in London, when a flood warning is received, the first action of the authorities is to send someone down to the river to check the level.

There was no self-checking (logic checking) in the software controlling the crane. It *believed* the crane was in a position where it could not open the hatch door while it was actually opening it! This is artificial stupidity, not artificial intelligence, greater stupidity than any human would have. No human operator would tell someone who could see him carrying out a hazardous task, "Don't worry, I am now somewhere else." However, in one plant (see Section 7.4), certain packages were supposed to be moved downstairs only in the elevator. When the elevator was out of use, no one noticed anything anomalous when a package arrived downstairs.

If it is essential to override a protective system, this should be signaled in a prominent way, for example, by a large red sign, not just by a message in a log book.

After the incident, the regulators asked the company to replace the software interlocks by hardwired ones. If safety interlocks are based on software, they should at least be on a separate PES from the normal control and operating system.

It is not possible to test software to confirm that every possible fault condition is covered but it does seem that interlock (c) was not adequately tested.

Because there were so many layers of protection, they were not given the highest safety rating and changes were not studied as thoroughly as they should have been. This was a common mode failure affecting all layers of protection. One strong wall around a castle is better than several weak walls. Several strong walls are better still.

In complex systems, it is difficult for people to understand all possible ramifications. Checklists can help them avoid errors but they are a poor substitute for simplicity.

Underlying these failures were weaknesses in management, in the training of the foreman and shift manager, in the control of modifications and permits-to-work (were they audited?), in the design of control systems and software, and, above all, in the belief that several poor systems make a good system. Replacing a weak system by an entirely new one is better than adding complexity.

The official report [7] on which this description is based aroused little interest outside the nuclear industry. Its title suggested that it was of local interest only. Titles, keywords, and abstracts of accident reports often ignore the lessons of major interest. They tell us the about the materials and equipment involved and the result, such as a leak, fire, or explosion, but do not tell us what we most need to know: the actions needed to prevent a recurrence, such as better control of modifications or better preparation for maintenance.

Some years later, when similar equipment was being designed elsewhere, someone asked if the design team had read the report on this incident. They had but the first reply was that the design standards for this type of equipment had been changed to incorporate the lessons learned. However, reading a standard does not have the same impact on the reader as reading an accident report. Designers should always look for reports on equipment or plants similar to those they are designing. Operating staff moving to a new process should read reports on past accidents on that and similar processes. Needless to say, such reports should be readily available. Section 2.7 describes an overcomplex manual system for a similar situation; it also failed.

14.7 "FAILURES IN MANAGEMENT, EQUIPMENT, AND CONTROL SYSTEMS"

The heading is taken from the official report [8]. So many things were wrong that this account could go in almost any chapter of this book. The incident occurred on the distillation section of a catalytic cracker but the messages are general. Please read on even if you have never seen a catalytic cracker or any oily equipment, and especially if you have ever:

- used computers for process control but not provided overview pages;
- made modifications to plants or processes but did not systematically consider possible consequences or did not provide training and instruction on how to operate after the change;
- installed more alarms than operators can possibly cope with;
- carried out insufficient inspection for corrosion or inspected in the wrong place;
- not learned, reviewed, and remembered the lessons of past experience on plants similar to your own; and
- reduced operator manning.

If you have never done any of the foregoing, perhaps your halo is obscuring your view.

The details of the incident are complex but were briefly as follows: A lightning strike caused a small fire on a catalytic cracker. During the resulting plant upset, the flow to a fractionation column was lost and the take-off valve on the bottom of the column closed automatically to maintain a level in the column. When the flow to the column was restarted, a light on the panel told the operators that this valve had reopened but it had not.

The liquid level in the column rose and the relief valve lifted and discharged some of the contents to a knock-out drum. The operators tried to reduce the level in the column by diverting some of the contents elsewhere but they reached the knock-out drum by another route. The drum filled and could not be emptied as its contents were automatically pumped back into the fractionation section for recovery. About 20 tons of liquids entered the pipe leading from the knock-out drum to the flare stack. The pipe was weakened by corrosion — its thickness was down to 0.3 mm — and it failed. The drifting cloud of vapor was ignited by the flare stack, 110 m (360 ft) away, producing an explosion, followed by a fire that was allowed to burn for two days as this was the safest way of disposing of the oil.

Damage was extensive but fortunately no one was killed or seriously injured. By good fortune, it was a Sunday afternoon so there were few personnel on site. A van carrying contractors was about to enter the area covered by the fireball but it was still a short distance away and a group of men had just left a building that was demolished.

14.7.1 Better Management Could Have Prevented the Incident

Over 4 h elapsed between the lightning strike and the explosion, thus some managers and day supervisors had come into the plant. However, instead of standing back, assessing the situation, and trying to diagnose what was happening, they got involved in hands-on operations (as in Section 4.1). It is not clear from the report whether or not they did so because there were too few operators to cope with the emergency.

The maintenance team had noticed that the flare line was corroding and thus had increased the frequency of inspections but not in the area where the pipe was thinnest, that is, near welds, especially longitudinal welds, as access was difficult at these places. This is rather like the story of the man who was seen after dark on his knees under a lamppost looking for something he had dropped. A passerby joined him and after a while asked if he was sure he had dropped it near the lamppost. "Probably not," said the man, "but the light is better here."

There was no system for reviewing, storing, or recalling incident information from similar plants. As shown in the following, the systems for the control of modifications, for the maintenance of instruments, and for monitoring corrosion were all flawed.

The company was prosecuted and had to pay $500,000 in fines and costs. The damage to the plant amounted to $75 million (at 1994 prices).

14.7.2 Better Control of Modifications Could Have Prevented the Incident

The pump-out system on the knock-out drum was modified a few years before the explosion so that the liquid in it was pumped back to the plant for reprocessing instead of going to slops. This meant that the liquid in it was pumped back to the system from which it had come so no reduction in the amount of liquid in the unit was achieved. It seems that when the modification was designed, no one foresaw that there might be a time when it would be necessary to reduce the total amount of liquid in the unit. They foresaw only a need to reduce the amount of liquid in individual vessels. A hazard and operability study might have disclosed the unforeseen result.

It was possible by valve operation to revert to the original design but this procedure had fallen into disuse from lack of practice and the absence of any written instructions.

14.7.3 Better Process Control Could Have Prevented the Incident

The failure of the column take-off valve was not an unfortunate and isolated incident. Thirty-nine control loops were examined afterward and 24 were found to be faulty. Many of the faults were known but repairs had been left to the next turnaround.

There were five product streams, with the flow rates spread across several display units. This made it difficult for the operators to assess the total output. There was no overview page.

The operators did not understand that some readings, such as temperatures and pressure, are based on direct measurements but others such as valve positions are based on indirect measurements. Thus the light that indicated that the take-off valve was open really showed that a signal had been sent to the valve telling it to open. It did not inform the operators whether or not the valve had actually opened. (A similar misunderstanding occurred at Three Mile Island [9].) Similarly, many pump running indicators are based on the voltage supplied to the pump motor. Whenever practical, we should measure directly what we want to know, not something from which it can be inferred.

There were far too many alarms for the operators to cope with, 755 out of 1,365 measurements had an alarm fitted while 431 had two alarms. At times, alarms were sounding every few seconds and operators were acknowledging them without realizing what they meant; 275 sounded in the last 11 min before the explosion. Records showed that the high-level alarm on the knock-out drum sounded 25 min before the explosion but it and the other critical alarms were acknowledged and overlooked. Safety critical alarms were not distinguished from others.

There was, of course, a case for each alarm if it was considered in isolation but we should never consider each problem on its own without also considering the total effect of our individual decisions.

14.8 CHANGES TO DESIGN AND OPERATIONS

Getting rid of waste products produces as many problems in plant design and operation as in the human body. As mentioned in Section 14.5.1, drain lines often have no line number and are overlooked in Hazop studies while sewers are often the problems of someone else.

A shallow pit, 0.5 m (20 in) deep, at a paper mill collected spillages from a tank truck offloading area. On the day of the incident, sodium hydrosulfide (NaSH) was being offloaded. As contractors were working nearby on a construction project, an operator drained the pit into the sewer. Unfortunately, the flow through the sewer was lower than usual and sulfuric acid was being added to the sewer to control its pH. The acid reacted with the NaSH to form hydrogen sulfide and within five minutes of the draining, the hydrogen sulfide had escaped through a leaking manway seal. Two contractors were killed and eight injured.

No management of change study was carried out when the pit was connected to the sewer — a change in design; nor when the addition of acid was started — a process change.

No one seems to have realized that mixing NaSH and acid would produce hydrogen sulfide, although the NaSH supplier's material safety data sheet stated that it formed hydrogen sulfide in contact with acid. As a result, there were no monitors, alarms, warning signs, or training on the action to take if it was formed.

The seal on the manway cover was known to be leaking, and repairs had been requested but never made. The leak had not been investigated though it was the company's policy to investigate near-misses.

Underlying these findings was the lack of an effective safety policy. There may have been a policy but if so, it was ignored in practice. The incident also shows the importance of giving as much detailed attention to drains and sewers, in both design, modification and operations, as to any other items of equipment [10].

14.9 THE IRRELEVANCE OF BLAME

The report on the last incident illustrates the truth of the following extract from an official UK report [11]:

The fact is — and we believe this to be widely recognized — the traditional concepts of the criminal law are not readily applicable to the majority of infringements that arise under this type of legislation. Relatively few offenses are clear cut, few arise from reckless indifference to the possibility of causing injury, and few can be laid without qualification at the door of a single individual. The typical infringement or combination of infringements arises rather through carelessness, oversight, lack of knowledge or means, inadequate supervision, or sheer inefficiency. In such circumstances prosecution and punishment by the criminal courts is largely irrelevant. The real need is for a constructive means of ensuring that practical improvements are made and preventative measures adopted.

This report was written in 1972 and led to major changes in UK legislation. Unfortunately, the report has been forgotten and there is an increasing tendency to look for culprits. The operator who closed the wrong valve is now less likely to be made the culprit; instead we now look higher up the management tree. There is still, however, a simplistic belief that someone must be to blame and a failure to realize that many people had an opportunity to prevent every accident. I have a lot of sympathy with the manager who wrote recently: "It is becoming increasingly hard to strike the right balance between the search for total safety and keeping [the plant running] . . . what really winds me up is the suggestion that people like me would put 'profit before safety' . . . As well as an insult to my integrity, I personally find it very offensive. I feel that I carry a weighty responsibility for the lives and livelihoods of the people who entrust themselves to our [operations]." [12]

REFERENCES

1. Crawley, F.K. (1998). Beware of steam condensing downstream of relief valves. *Loss Prevention Bulletin*, Oct., 143:19–20.

2. Gillard, T. (1998). Seizure of boiler relief valves. *Loss Prevention Bulletin*, June, 141:14–15.

3. Anon. (1998). *Investigation Report: Catastrophic Vessel Overpressure*. Report No. 1998-002-I-LA, US Chemical Safety and Hazard Investigation Board, Washington, D.C.

4. Kletz, T.A. (2001). *Learning from Accidents*, 3rd ed., Butterworth-Heinemann, Oxford, UK and Woburn, MA, Chapter 8.

5. Anon. (1958). *Hazards of Air*, American Oil Company, Chicago, IL, and later editions, pp. 32–33.

6. Fishwick, T. (1998). Three flare stack incidents. *Loss Prevention Bulletin*, Aug., 142:7–11.

7. HM Nuclear Installations Inspectorate (1992). *Windscale Vitrification Plant Shield Door Incident*, Her Majesty's Stationery Office, London.

8. Health and Safety Executive (1997). *The Explosion and Fire at the Texaco Refinery, Milford Haven, 24 July 1994*, HSE Books, Sudbury, UK.

9. Kletz, T.A. (2001). *Learning from Accidents*, 3rd ed., Butterworth-Heinemann, Oxford, UK and Woburn, MA, Chapter 11.

10. Anon. (2003). *Investigation Report — Hydrogen Sulfide Poisoning*. Report No. 2002-01-1-AL, US Chemical Safety and Hazard Investigation Board, Jan.

11. Anon. (1972). *Safety and Health at Work — Report of the Committee (the Robens Report)*, Her Majesty's Stationery Office, London, paragraph 26.1.

12. Holden, M. (2001). *Modern Railways*, Jan., **58**(628):37.

Chapter 15

Accidents in Other Industries

During an Alpine hike in 1948, Swiss mountaineer George de Mestral became frustrated by the burs that clung annoyingly to his pants and socks. While picking them off, he realized that it might be possible to produce a fastener based on the burs to compete with, if not obsolete, the zipper.

— Charles Panati, *Extraordinary Origins of Everyday Things*

As this quotation shows, new ideas often result when we transfer an idea from one field of knowledge to another. In the same way, we can learn from accidents in other industries. Their immediate technical causes are not always of interest but the underlying causes can supplement and reinforce our own experience. Because we are not involved in the technical details, we often see the underlying causes more clearly and reading about them is more recreation than it is work. And it may also be comforting to learn that people in other industries make as many errors as we do.

15.1 AN EXPLOSION IN A COAL MINE

An explosion in a Canadian coal mine in 1992 killed 27 people and led to the bankruptcy of the parent company. There was an explosion of methane, which set off a coal dust explosion. The source of ignition was probably sparks formed by mining machinery striking rock. This was a triggering event rather than a cause as there were so many ongoing faults that an explosion in the end was almost inevitable.

- Inadequate ventilation allowed explosive mixtures of methane and air to form.
- The methods for detecting methane were also inadequate and mining was allowed to continue when such methods were inoperable.
- Too much coal dust was allowed to accumulate.
- It is normal practice to dilute coal dust with stone dust to prevent explosions but not enough stone dust was used as stocks were too low.
- Many of the miners were inexperienced and inadequately trained; 12-h shifts made them tired; fear of reprisals discouraged the reporting of hazards and those that were reported were not followed up.
- Output was put before safety.

There were other hazards not directly connected with the explosion. Thus falls of roof were common as intersecting fault lines were ignored [1].

These various shortcomings were not, of course, isolated. They all stemmed from a cavalier attitude to safety at all levels but particularly among senior managers as they set the example that others will follow.

This plant was so bad that many readers may feel that it has no message for them. But good plants can deteriorate, perhaps after a change of management (see Chapter 4). Stocks of spares are reduced to save cost, instruments are tested or maintained less frequently, and before long the plant has started going downhill. Every journey, uphill or downhill, starts with one step. Sections 8.4 and 8.9 describe other dust explosions.

15.2 MARINE ACCIDENTS

Many marine accidents are process accidents, similar to many that have occurred, or could occur, in chemical plants. The following are all taken for the periodic reports published by the UK Marine Accidents Investigation Branch.

15.2.1 A Misleading Display

A roll-on, roll-off ferry was ready to depart. When one of the ship's officers tried to close the bow door (the type that lifts up like a knight's visor), it refused to move. He sent for the engineers who decided to close the door manually. They stopped immediately when they realized that the door was buckling. It was then discovered that a bolt that held the door open was still engaged. There was a light on the control panel next to a *Visor open* label and the operators assumed that this meant that the visor was free to be lowered. The light actually referred to something quite different. The report [2] says:

> The layout of any control panel must be clear and unambiguous. . . . If it is capable of being read wrongly, you can be sure it will! Crews come and go, and unless instructions are up to date and clear and easily understood, experience and word-of-mouth explanations get lost.

Visual checks of the locking mechanism were difficult and time-consuming. After the incident, this locking mechanism was changed to make inspection easier. As in so many of the incidents described in this book, what looks at first sight as poor operations could have been prevented, or made less likely, by better design. Sometimes fundamental redesign is needed but often, as in this case, all we need is more attention to detail.

15.2.2 Stand Clear

When we are carrying out a pressure test, we know that the equipment we are testing might fail and so people should always position themselves to prevent injury if it does. Failures during pressure testing are rare but not unknown. If we were sure the equipment would not fail, we would not need to test it (though testing, as well as proving that the equipment is safe to use at its design pressure, also relieves stress).

In the same way, when moving machinery is started up for the first time after repair, we should remember that it might fail. A centrifugal lubrication oil purifier on a ship had been reassembled after maintenance. It was run empty without trouble but when oil was admitted the bowl burst. Fortunately, only a minor injury was incurred.

The failure was due to an error in assembly. There was a change of shift during the assembly and it seems that the second shift misunderstood exactly what still needed to be done, or were not adequately briefed [3]. Equipment should be designed so that it cannot be assembled incorrectly.

A cooling water pump on a passenger ferry failed and one of the engines overheated. The first component to fail was the exhaust gas trunking, which started to melt. This allowed exhaust gases to percolate into the passenger areas. One of the passengers noticed the fumes and reported this to the crew. What, the report wonders, might have happened if he had not raised the alarm at an early stage? [4]

There was a high-temperature alarm on the engine but it did not operate. Perhaps it was out of order or perhaps the set point was fixed to protect the engine and no one realized that the trunking was more vulnerable. A similar incident occurred on a chemical plant. An electric heater was fitted with a high-temperature trip. The set point was chosen to prevent damage to the heater elements but no one realized that the body of the heater would be damaged at a lower temperature. In fact, it ruptured.

15.2.3 Wrong Connections

A fishing vessel left port after an overhaul. Soon afterward the bearings on the engine turbocharger seized. Fortunately, the small fire that followed was soon extinguished and the vessel was towed back to port. It was then found that during reassembly, the bearing oil supply had been connected to the cooling water inlet and the cooling water supply to the bearing oil inlet.

It is easy to say, as the report [5] does, that the ship's crews should always check the work of contractors. Of course they should, but accidents such as this will continue to occur until designers learn to use different types or sizes of connections for different duties. In the meantime, users should paint different connections different colors (or attach colored sticky tape to them).

15.2.4 Preparation for Maintenance

Many accidents have occurred in industry because maintenance was undertaken without adequate consideration of the risk (see Chapter 1). The same is true at sea. For example, a rising engine temperature showed that the sea water inlet was blocked. The mate closed the seacock as far as he could and then removed the cover from the box on the ship's side of the seacock. He removed a plastic bag, which had been sucked into it. As he did so, water started to pour out. He tried to close the seacock further but broke the linkage. The bilge pumps could not cope with the flow, and the ship was abandoned and sank a few hours later [6].

15.2.5 Entry into Confined Spaces

Many people have been killed or overcome because they entered tanks or other confined spaces on ships without authorization or before the atmosphere had been tested. Sometimes the procedures were poor but often they were ignored (see Chapter 2). The following incident is more unusual.

Frozen fish was being loaded into a refrigerated ship. Once on board, the pallets were stowed by LPG-powered forklift trucks in enclosed decks (the tween decks). The stevedores complained of headaches and nausea and loading was stopped. The cause was a build-up of carbon monoxide. Most people know that internal combustion engines should not be operated in a confined space, such as a garage. In this case, the low temperature ($-20\,°C$, $-4\,°F$) reduced the effectiveness of combustion and led to increased production of carbon monoxide, oxides of nitrogen and unburnt fuel.

Electrically powered vehicles should be used in confined spaces. This is stated in the UK Code of Safe Working Practice for Merchant Seamen but seems to have been unknown to the ship's officers and the stevedores' employer [7].

15.2.6 For Want of a Nail a Ship Was Lost

Two UK fishing vessels sank within 10 days of each other. Fortunately, the crews were rescued. In both cases the seawater pipes leaked, probably as the result of corrosion, and flooded the engine rooms. Both ships were fitted with water-level alarms that failed to work, probably because they were not tested regularly, if at all, but possibly because the wiring was not in protective conduits. By the time the flooding was discovered, it was too late to close the valves in the seawater lines as they were below the water level. Extended spindles on the valves would have saved the ships [8].

Maintenance and operations on small ships (and small plants?) are often poor but nevertheless we can learn from these incidents. Who has never postponed the testing of alarms and trips because the testers were too busy elsewhere? Who has never overlooked the opportunity to make a cheap change that would add an extra layer of protection? Valves that are normally left open or shut but that might be needed when things go wrong should be operated regularly, say, every week, so that they do not become stiff. Are yours?

15.3 HUMAN ERROR

On ships, as on land, there is a readiness to blame human error —
poor maintenance, watchkeepers falling asleep, errors in navigation —
instead of looking for underlying causes such as poor training or
supervision, error-prone designs, lack of protective features, overlong
hours of work, and so on. In the following extract from a report [9] on
marine accidents, I have changed *seamen* to *operators* and made other
similar changes:

> There is an abundance of academic literature on human error which
> quickly lapses into language that leaves the average operator [and
> engineer] totally bewildered, and few will have the foggiest ideas
> what is meant by "visual/tactile dissimilarity," "cognitive aspects of
> safety," "rule-based behaviour," "latent conditions and pathogens," or
> "non-optimised performance related factors." What the operator [and
> the engineer] needs is a simple explanation about what is meant by
> human factors so he or she can better understand why it matters and
> what needs to be done to improve safety and conditions of service.

I have tried to provide such a guide in *An Engineer's Guide to Human
Error* [10] (see the Introduction and Section 16.3).

The following are two more adapted quotations from a marine report
[11], this time without change:

> [When a ship has run aground] giving orders calmly will ensure
> success. It is not the moment to give the unfortunate helmsman his
> or her annual appraisal.
>
> When the draught of your vessel exceeds the depth of water avail-
> able . . . you can always consider the delights of gardening.

Section 4.1 drew attention to the reluctance of some operators to go and
look at the plant when it is not operating correctly. The following extract
from a letter by a deep sea pilot [12] describes a view shared by many
chemical engineers:

> Modern watch-keepers tend to be wonderful at operating computers
> and twiddling radars, but abysmal in the basics, such as keeping a
> visual lookout and correctly applying the collision regulations. Lack
> of a grounding in mental arithmetic also means that they often cannot

roughly estimate their computerized information and realize when it is wrong.

There are other marine accident reports in Sections 5.5 and 7.3.

15.4 TESTS SHOULD BE LIKE REAL LIFE

Section 14.1 in *WWW* describes several tests that did not detect faults because they did not simulate real-life conditions, for example, a high-temperature trip was removed from its case before testing. The tests therefore failed to detect that the pointer was touching the plastic front of the instrument case and this prevented it from reaching the trip point. The following is an example from another industry.

To the surprise of the manufacturers, a small car failed to pass a rear collision test. It crumpled more than expected. It was then found that the testers had removed the spare wheel before the test as it seemed unnecessarily wasteful to damage it. However, the spare wheel, correctly inflated, was a necessary part of the energy-absorbing process [13].

15.5 LOAD AND STRENGTH TOO CLOSE

As described in Section 3.3.2, in 2000 a railway accident at Hatfield, UK killed four people and drew attention to a literal interface problem. The immediate cause was a cracked rail but the underlying cause was that British Railways had been privatized and split into a hundred companies. Responsibility for the rails and the wheels now lay with different organizations. To quote the head of the railways' safety organization, "Both sides of the wheel/rail interface may be operating within their respective safety based Standards, but the combined effect of barely acceptable wheel on barely acceptable rails is unacceptable" [14].

Figure 15-1 may make this clearer. In any system, the strength and the load vary to some extent from their design values and there is inevitably a small overlap between the two asymptotes. Its area is a measure of the probability that the load will exceed the strength and the system will fail, not necessarily immediately but in the long run. Normally, this probability is negligible. In the case of the railways, the wear on the wheel increases the load and cracks in the rail decrease the strength. Both were just within specification and the overlap was too large. This led to rolling contact fatigue of the track (also called gauge corner cracking), the train

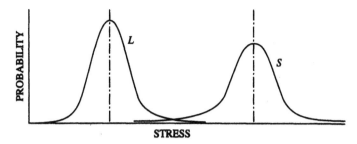

Figure 15-1. Overlapping distributions of load (L) and strength (S). Reprinted with permission of *H & H*, Fig. 7.1, 1999, p. 208.

crash, and the consequent upheaval while hundreds of miles of faulty rail were replaced.

The engineering principle involved is hardly new. In 1880, Chaplin showed that a chain can fail if its strength is at its lower limit and the load is at its upper limit [15]. The Hatfield crash did not occur because engineers had forgotten this but because there were no engineers in the senior management of the company (Railtrack) that owned the track. They had all been moved to the maintenance companies (or elsewhere) and Railtrack had lost the knowledge that it needed to make it an intelligent customer of the maintenance companies. The change had a further disadvantage:

> As one senior railwayman put it: In an integrated railway you could gain a lot more skills; you could work through managing train crews, signalling, running stations — you really got a feel for everything in the business, you would learn about every activity so that you knew how the railways operated.

> Now with so many employees following paths through a single company, their experience is so much more limited in that the broad base of knowledge has disappeared. The steady flow of skilled operators and skilled engineers ready to take up senior management positions has created a dire shortage.

> As an example, in the years prior to 1994, the railways took on an average of 40 engineering graduates annually, and then between 1994 and 2002, the total was almost zero [16].

REFERENCES

1. Amyotte, A.M. and A.M. Oehmne (2002). Application of a loss causation model to the Westray mine explosion, *Process Safety and Environmental Protection*, **80**(B1):55.

2. Anon. (2001). Light reading! *Safety Digest — Lessons from Marine Accident Reports*, No. 1/2001, Marine Accident Investigation Branch of the UK Department of Transport, Local Government and the Regions, London, p. 13.

3. Anon. (2001). The explosive force of high speed rotation! *Safety Digest — Lessons from Marine Accident Reports*, No. 3/2001, Marine Accident Investigation Branch of the UK Department of Transport, Local Government and the Regions, London, p. 20.

4. Anon. (2002). Melting moments, *Safety Digest — Lessons from Marine Accident Reports*, No. 2/2002, Marine Accident Investigation Branch of the UK Department of Transport, London, p. 13.

5. Anon. (2002). The thinking man, *Safety Digest — Lessons from Marine Accident Reports*, No. 2/2002, Marine Accident Investigation Branch of the UK Department of Transport, London, pp. 36–37.

6. Anon. (2002). Ideas for tackling flooding, *Safety Digest — Lessons from Marine Accident Reports*, No. 2/2002, Marine Accident Investigation Branch of the UK Department of Transport, London, pp. 43–44.

7. Anon. (2001). Cold comfort, *Safety Digest — Lessons from Marine Accident Reports*, No. 3/2001, Marine Accident Investigation Branch of the UK Department of Transport, Local Government and the Regions, London, pp.18–19.

8. Anon. (2001). Defective bilge alarms lead to the loss of two large vessels, *Safety Digest — Lessons from Marine Accident Reports*, No. 2/2001, Marine Accident Investigation Branch of the UK Department of Transport, Local Government and the Regions, London, pp. 46–47.

9. Anon. (2001). A pause for thought, *Safety Digest — Lessons from Marine Accident Reports*, No. 3/2001, Marine Accident Investigation Branch of the UK Department of Transport, Local Government and the Regions, London, pp. 50–52.

10. Kletz, T. A. (2001). *An Engineer's View of Human Error*, 3rd ed., Institution of Chemical Engineers, Rugby, UK and Taylor and Francis, Philadelphia. PA.

11. Anon. (2001). When the draught of your vessel exceeds the depth of water ..., *Safety Digest — Lessons from Marine Accident Reports*, No. 2/2001, Marine Accident Investigation Branch of the UK Department of Transport, Local Government and the Regions, London, pp. 60–64.

12. Francis, R. (2002). Declining standards on the watch. *Daily Telegraph* (London), 18 Dec., p. 21.

13. Ford, R. (1998). Crashworthiness: an ethical issue. *Modern Railways*, Sept., **55**(600):576–577.

14. Ford, R. (2002). Gauge corner cracking — privatisation indicted. *Modern Railways*, Jan., **59**(640):19–20.

15. Pugsley, A.G. (1966). The Safety of Structures, Arnold, London (quoted by N.R.S. Tait (1987). *Endeavour*, **11**(4):192).

16. Miles, T. (2003). Where are the aspiring managers? *Modern Railways*, Feb., **60**(653):65.

Chapter 16

Accident Investigation — Missed Opportunities

If the origin of the human mind is to be understood, it is important to be able to identify signals of distinctly non-human behaviour. Lack of innovation is one of them.

— Roger Lewis, *The Origin of Modern Humans*

Almost all the accidents described in this book need not have occurred. Similar ones have happened before and accounts of them have been published. Someone knew how to prevent them even if the people on the job at the time did not. This suggests that there is something seriously wrong with our accident investigations, safety training, and the availability of information.

Having paid the price of an accident, minor or serious (or narrowly missed), we should use the opportunity to learn from it. Failures should be seen as educational experiences. The 10 major opportunities summarized in what follows, are frequently missed, the first 7 during the preparation of a report and the other 3 afterward. Having paid the *tuition fee*, we should learn the lessons. The evidence is usually collected adequately; the weakness lies in its interpretation.

16.1 ACCIDENT INVESTIGATIONS OFTEN FIND ONLY A SINGLE CAUSE

Often, accident reports identify only a single cause, though many people, from the designers, down to the last link in the chain, the mechanic who broke the wrong joint or the operator who closed the wrong valve, had an opportunity to prevent the accident. The single cause identified is usually this last link in the chain of events that led to the accident. Just as we are blind to all but one of the octaves in the electromagnetic spectrum, we are blind to many of the opportunities that we have to prevent an accident. But just as we have found ways of making the rest of the spectrum visible, we need to make all the ways of preventing an accident visible.

16.2 ACCIDENT INVESTIGATIONS ARE OFTEN SUPERFICIAL

Even when we find more than one cause, we often find only the immediate causes. We should look beyond them for ways of avoiding the hazards, such as inherently safer design. For example, could less hazardous raw materials have been used? Also, we should look for weaknesses in the management system. For example, could more safety features have been included in the design? Were the operators adequately trained and instructed? If a mechanic opened up the wrong piece of equipment, could there have been a better system for identifying it? Were previous incidents overlooked because the results were, by good fortune, only trivial? The emphasis should shift from blaming the operator to removing opportunities for error or identifying weaknesses in the design and management systems.

Most of the chapter headings in this book are examples of root causes and, as mentioned in the Preface, this has made the allocation of incidents to chapters somewhat arbitrary as most of them have more than one root cause.

When investigators are asked to look for underlying or root causes, some of them simply call the causes they have found root causes (see Section 12.3 for an example). One report quoted corrosion as the root cause of equipment failure but it is an immediate cause. To find the true root causes, we need to ask if corrosion was foreseen during design and if not, why not; were operating conditions the same as those given to the designer and if not, why not; was regular examination for corrosion requested, and if so, had it been carried out and the results acted upon, and so on. Senior managers should not accept accident reports that deal only with immediate causes.

The causes listed in accident reports sometimes tell us more about the investigators' beliefs and background than about the accidents. One company had recognized that failure to learn from past experience was a major cause of accidents and was making strenuous efforts to improve its learning from experience. However, none of their accident reports or the annual summary of them mentioned this as a cause. The members of the investigating panels did not know that similar accidents had happened before.

16.3 ACCIDENT INVESTIGATIONS LIST HUMAN ERROR AS A CAUSE

As mentioned in the Introduction, human error is far too vague a term to be useful. We should ask, "What sort of error?" because different sorts of error require different actions if we are going to prevent the errors from happening again [1].

- Was the error a mistake, that is, one due to poor training or instructions, so that the intention was wrong. We need to improve the training and instructions and, if possible, simplify the task. While instructions tell us what to do, training gives us the understanding that allows us to handle unforeseen situations. However many instructions we write, we will never foresee everything that might go wrong. (For examples see Sections 2.5.2, 7.4, 7.5, 8.12, and 14.5.)

- Was the error due to a violation or noncompliance, that is, a deliberate decision not to follow instructions or recognized good practice? If so, we need to explain the reasons for them as we do not live in a society in which people will simply do what they are told. We should, if possible, simplify the task — if an incorrect method is easier than the correct one, it is difficult to persuade everyone to use the correct method — and we should check from time to time to see that instructions are being followed.

- Was the task beyond the ability of the person asked to do it, perhaps beyond anyone's ability? If so, we need to redesign the task.

- Was it a slip or lapse of attention (like many of those described in Chapter 5). In contrast to mistakes, the intention may have been correct but it was not fulfilled. It is no use telling people to be more careful as no one is deliberately careless. We should remove opportunities for error by changing the design or method of working.

Designers, supervisors, and managers make errors of all these types though slips and lapses of attention by designers and managers are rare as they usually have time to check their work. Errors by designers produce traps into which operators fall, that is, they produce situations in which slips or lapses of attention, inevitable from time to time, result in accidents. Errors by managers are signposts pointing in the wrong directions.

16.4 ACCIDENT REPORTS LOOK FOR PEOPLE TO BLAME

In every walk of life, when things go wrong the default action of many people is to ask who is to blame? The banner headline in my newspaper after a railway accident was "Who is to blame this time?" However, blaming human error for an accident diverts attention from what can be done by better design or methods of operation. To quote James Reason, "We cannot change the human condition but we can change the conditions in which humans work." Even when people ask, "What did we do wrong?" they often find the wrong answer. They find that the instructions were perhaps not clear enough, rewrite them in greater detail and at greater length, and thus reduce the probability that anyone will read them. They should consider the alternative actions listed in Section 16.6 in what follows.

To paraphrase G. K. Chesterton, The horrible thing about all the people who work at plants, even the best, is not that they are wicked, not that they are stupid, it is simply that they have gotten used to it. They do not see the hazards; all they see is the usual people carrying out the usual tasks in the usual place. They do not see the risks; they see only their own place of work.

One method of jerking people out of their familiarity is to show them slides of the hazards they pass everyday without noticing them. On one occasion, I led a discussion of a leak that had occurred from a substandard drain point. Immediately afterward one of the people who had been present went into a compressor building that he visited every day. As he walked through the door, he saw a substandard drain point.

16.5 ACCIDENT REPORTS LIST CAUSES THAT ARE DIFFICULT OR IMPOSSIBLE TO REMOVE

For example, a source of ignition is often listed as the cause of a fire or explosion. But it is impossible on the industrial scale to eliminate all sources of ignition with 100% certainty. While we try to remove as many as possible, it is more important to prevent the formation of flammable mixtures.

Which is the more dangerous action on a plant that handles flammable liquids: to bring in a box of matches or to bring in a bucket? Many people would say that it is more dangerous to bring in the matches, but nobody would knowingly strike them in the presence of a leak and in a well-run plant leaks are small and infrequent. If a bucket is allowed in, however, it may be used for collecting drips or taking samples. A flammable mixture will be present above the surface of the liquid and may be ignited by a stray source of ignition. Of the two *causes* of the subsequent fire, the bucket is the easier to avoid.

I am not, of course, suggesting that we allowed unrestricted use of matches on our plants but I do suggest that we keep out open containers as thoroughly as we keep out matches. Instead of listing causes, we should list the actions needed to prevent a recurrence. This forces people to ask if and how each so-called cause can be prevented in future.

16.6 WE CHANGE PROCEDURES RATHER THAN DESIGNS

As discussed in Chapter 5, when making recommendations to prevent an accident, our first choice should be to see if we can remove the hazard — the inherently safer approach. For example, could we use a nonflammable solvent instead of a flammable one? Even if is impossible at the existing plant, we should note it for the future.

The second best choice is to control the hazard with protective equipment, preferably passive equipment, as it does not have to be switched on. As a last (but frequent) resort, we may have to depend on procedures. Thus, as a protection against fire, if we cannot use nonflammable materials, insulation (passive) is usually better than water spray turned on automatically (active), but that is usually better than water spray turned on by people (procedural). In some companies, however, the default action is to consider a change in procedures first, sometimes because it is cheaper but more often because it has become a custom and practice carried on unthinkingly.

Operators provide the last line of defense against errors by designers and managers. It is a bad strategy to rely on the last line of defense and to neglect the outer ones. Good loss prevention starts far from the top event, in the early stages of design. Blaming users is a camouflage for poor design.

16.7 WE MAY GO TOO FAR

Sometimes after an accident, people go too far and spend time and money on making sure that nothing similar could possibly happen again even though the probability is extremely unlikely. If the accident was a serious one, it may be necessary to do this to reassure employees and the public, but otherwise we should remember that if we goldplate one unit there are fewer resources available to silverplate the others.

As mentioned in Chapter 5, in the UK the law does not require companies to do everything possible to prevent an accident, only what is *reasonably practicable*. This legal phrase means that the size of a risk should be compared with the cost of removing it, in money, time, and trouble, and if there is a gross disproportion between them, it is not necessary to remove the risk. In recent years, the regulator, the Health and Safety Executive, has provided detailed advice on the risks that are tolerable and the costs that are considered disproportionate [2]. In most other countries, the law is more rigid and, in theory, expects companies to remove all risks. This, of course, is impossible but it makes companies reluctant to admit that there is a limit to what they, and society, can afford to spend even to save a life. (If this sounds cold-blooded, remember that we are discussing very low probabilities of death where further expenditure will make the probability even lower but is very unlikely to actually prevent any death or even injury.)

16.8 WE DO NOT LET OTHERS LEARN FROM OUR EXPERIENCE

Many companies restrict the circulation of incident reports, as they do not want everyone, even everyone in the company, to know that they have blundered. However, this will not prevent the incident from happening again. We should circulate the essential messages widely, in the company and elsewhere, so that others can learn from them, for several reasons as follows.

- Moral: if we have information that might prevent another accident, we have a duty to pass it on.
- Pragmatic: if we tell other organizations about our accidents, they may tell us about theirs.
- Economic: we would like our competitors to spend as much as we do on safety.

• The industry is one: every accident affects its reputation. To misquote the well-known words of John Donne:

No plant is an Island, entire of itself; every plant is a piece of the Continent, a part of the main. Any plant's loss diminishes us, because we are involved in the Industry: and therefore never send to know for whom the Inquiry sitteth; it sitteth for thee.

When information is published, people do not always learn from it. A belief that *our problems are different* is a common failing (see Section 12.3).

16.9 WE READ OR RECEIVE ONLY OVERVIEWS

This opportunity is one missed by many senior people. Lacking the time to read accident reports in detail they consume predigested summaries of them, full of generalizations such as, *There has been an increase in accidents due to inadequate training*. However, as already mentioned, the identification of underlying causes can be very subjective and is influenced by people's experience, interests, blind spots and prejudices. Senior people should read a number of accident reports regularly and, if necessary discuss them with their authors to see if they agree with the assignment of underlying causes. In any field of study, reliance on secondary sources instead of primary ones can perpetuate errors.

16.10 WE FORGET THE LESSONS LEARNED AND ALLOW THE ACCIDENT TO HAPPEN AGAIN

Even when we prepare a good report and circulate it widely, all too often it is read, filed, and forgotten. Every chapter shows that organizations have no memory [3]. Only people have memories and after a few years they move on, taking their memories with them. Procedures introduced after an accident are allowed to lapse and some years later, the accident happens again, even on the plant where it happened before. If by good fortune the results of an accident are not serious, the lessons are forgotten even more quickly (see Section 3.4).

The following are some actions that can prevent the same accidents from recurring so often:

• Include in every instruction, code, and standard a note on the reasons for it and accounts of accidents that would not have occurred if the instruction, etc. had existed at the time and had been followed. Once we forget the origins of our practices, they become *cut flowers*; severed from their roots they wither and die.

- Never remove equipment before we know why it was installed. Never abandon a procedure before we know why it was adopted.

- Describe old accidents as well as recent ones, other companies' accidents as well as our own, in safety bulletins and discuss them at safety meetings.

- Follow up at regular intervals to see that the recommendations made after accidents are being followed, in design as well as operations.

- Remember that the first step down the road to an accident occurs when someone turns a blind eye to a missing blind.

- Include important accidents of the past in the training of undergraduates and company employees.

- Keep a folder of old accident reports in every control room. It should be compulsory reading for recruits and others should look through it from time to time.

- Read more books, which tell us what is old, as well as magazines that tell us what is new.

- We cannot stop downsizing but we should make sure that the remaining employees at all levels have adequate knowledge and experience.

- Devise better retrieval systems so that we can find details of past accidents in our own and other companies more easily than at present, and the recommendations made afterward. We need systems in which the computer will automatically draw our attention to information that is relevant to what we are typing or reading (see Section 16.10.2).

- Everyone forgets the past. An historian of football found that fans would condense the first hundred years of their team's history into two sentences and then describe the last few seasons in painstaking detail. (But engineers' poor memories have more serious results.)

16.10.1 Weaknesses in Safety Training

There is something seriously wrong with our safety education when so many accidents repeat themselves so often. (Speaking at a conference on the lessons of Three Mile Island, Norman Rasmussen said that "we do a lot of teaching, it's just that we don't get much learning done in some of these schools" (4).) The first weakness is that IT IS OFTEN TOO THEORETICAL. It starts with principles, codes, and standards. It tells us what we should do and why we should do it and warns us that we may have accidents if we do not follow the advice. If anyone is still reading or listening, it may then go on to describe some of the accidents.

We should start by describing accidents and draw the lessons from them for two reasons. First, accidents grab our attention and make us read on, or sit up and listen. Suppose an article describes a management system for the control of plant and process modifications. We probably glance at it and put it aside to read later, and you know what that means. If it is a talk, we may yawn and think, *Another management system designed by the safety department that the people at the plant will not follow once the novelty wears off.* In contrast, if someone describes accidents caused by modifications made without sufficient thought, we are more likely to read on or listen and consider how we might prevent them in the plants under our control. We remember stories about accidents far better than we remember disconnected advice.

Whatever the subject, we should build generalities from individual cases; otherwise they have no foundations.

The second reason why we should start with accident reports is that the accident tells us what actually happened. You may not agree with my recommendations but I hope you will not ignore the events I have described. If they could happen at your plant, I hope you will take steps to prevent them, though not necessarily the steps that I have suggested.

A second weakness with our safety training is that it usually consists of talking to people rather than discussing safety training with them. Instead of describing an accident and the recommendations made afterward, outline the story and let the audience question you to find out the rest of the facts, those that they think are important and that they want to know. Then let them say what THEY THINK ought to be done to prevent it happening again. More will be remembered and the audience will be more committed than if they were merely told what to do.

Once someone has blown up a plant, they rarely do so again, at least not in the same way. But when he or she leaves, the successor lacks the experience. Discussing accidents is not as effective a learning experience as letting them happen but it is the best simulation available and it is a lot better than reading a report or listening to a talk.

We should choose for discussion accidents that bring out important messages such as the need to look for underlying causes, the need to control modifications, the need to avoid hazards rather than to control them, and so on. You can discuss the accidents described in this book but it would be better to discuss those that occurred in your own plant. The audience cannot then think, *We would not do anything as stupid as the people at that plant.*

Undergraduate training should include discussion of some accidents, chosen because they illustrate important safety principles. If universities do not provide this sort of training, industry should provide it. In any case, new recruits need training on the specific hazards of the industry.

16.10.2 Databases

Accident databases should, in theory, keep the memory of past incidents alive and prevent repetitions, but they have been used less than expected. A major reason is that we check a database only when we suspect that there might be a hazard. If we do not suspect there may be a hazard, we do not check.

In conventional searching, the computer is passive and the user is active. The user has to ask the database if there is any information on, say, accidents involving particular substances, operations, or equipment. The user has to suspect that there may be a hazard or he or she will not check. We need a system in which the user is passive and the computer is active. With such a system, if someone is using a word processor, a design program, or a Hazop recording program and types "X," the computer will signal that the database contains information on this substance, subject, or equipment. A click of the mouse will then display the data. As I type these words, the spellcheck and grammar check programs are running in the background drawing my attention to my (frequent) spelling and grammar errors. In a similar way, a safety database could draw attention to any subject on which it has data. Filters could prevent it repeatedly referring to the same hazard [5].

A program of this type has been developed for medical use. Without the doctor taking any action, the program reviews the information on symptoms, treatment, diagnosis, etc. already entered for other purposes and suggests treatments that the doctor may have overlooked or not be aware of.

When we are aware that there is or may be a hazard and carry out conventional searching, it is hindered by another weakness: it is hit or miss. We either get a *hit* or not. Suppose we are looking in a safety database to see if there are any reports on accidents involving the transport of sulfuric acid. Most search engines will display them or tell us there are none. A *fuzzy* search engine will offer us reports on the transport of other minerals, acids, or perhaps on the storage of sulfuric acid. This is done by arranging keywords in a sort of family tree. If there are no reports on the keyword, the system will offer reports on its parents or siblings [6,7].

16.10.3 Cultural and Psychological Blocks

Perhaps there are cultural and psychological blocks, which encourage us to forget the lessons of the past.

- We live in a society that values the new more than the old, probably the first society to do so. Old used to imply enduring value, whether applied to an article, a practice, or knowledge. Anything old had to be good to have lasted so long. Now it suggests obsolete or at least obsolescent.

- We find it difficult to change old beliefs and ways of thinking.

- A psychological block is that life is easier to bear if we can forget the errors we have made in the past. Perhaps we are programmed to do so.

The first step towards overcoming these blocks is to realize that they exist and that engineering requires a different approach. We should teach people that "It is the success of engineering which holds back the growth of engineering knowledge, and its failures which provide the seeds for its future development" [8].

REFERENCES

1. Kletz, T.A. (2001). *An Engineer's View of Human Error*, 3rd ed., Institution of Chemical Engineers, Rugby UK and Taylor & Francis, New York.

2. Health and Safety Executive (2001). *Reducing Risks, Protecting People — HSE's Decision Making Process*, HSE Books, Sudbury, UK.

3. Kletz, T.A. (1993). *Lessons from Disaster — How Organisations have No Memory and Accidents Recur*, Institution of Chemical Engineers, Rugby UK and Gulf, Houston, TX.

4. Rasmussen, N.C. General Discussion in T.H. Moss, and D.L. Sills (eds.), (1999). *The Three Mile Island Nuclear Accident — Lessons and Implications*, New York Academy of Sciences, New York, NY, p. 50.

5. Bond, J. (2003). Linking an accident database to design and operational software, *Hazards XVII: Process Safety — Fulfilling our Responsibilities, Symposium Series No. 149*, Institution of Chemical Engineers, Rugby, UK, pp. 491–500.

6. Chung, P.W.H. and M. Jefferson (1998). A fuzzy approach to accessing accident databases. *Applied Intelligence*, 9:129–137.

7. Iliffe, R.E., P.W.H. Chung, and T.A. Kletz (1998). *Hierarchical Indexing, Some Lessons from Indexing Incident Databases*, International Seminar on Accident Databases as a Management Tool, Antwerp, Belgium, Nov.

8. Blockley, D.I. and J.R. Henderson (1980). *Proceedings of the Institution of Civil Engineers*, Part 1, 68:719.

SOME TIPS FOR ACCIDENT INVESTIGATORS

DO NOT SET A TARGET FOR DANGEROUS INCIDENTS. If you do, people will find reasons why some should not be counted and the target will always be met.

DO NOT LOOK FOR CULPRITS TO BLAME. Today, everybody says they do not, but after an accident many revert to old ways of thinking.

AN INDULGENT ATTITUDE TO NONCOMPLIANCE IS USUALLY A PRICE WORTH PAYING TO FIND OUT WHAT REALLY HAPPENED. Remember that many violations occur because people are trying to help; they think they have found a better way of carrying out a task.

TO FIND OUT WHAT HAS NOT BEEN REPORTED, KEEP YOUR EYES AND EARS OPEN AND *LUNCH AROUND*, that is, do not lunch with the same people every day. If you are asked to approve claims for damaged clothing or overtime for cleaning up spillages, ask if the incident has been investigated and reported.

ALWAYS VISIT THE SITE OF ACCIDENTS and look where others do not, behind and underneath equipment. Look at neighboring areas for comparison.

PHOTOGRAPH THE SCENE for inclusion in the report — a photograph may tell us more than a thousand words — and for future use in safety courses and publications.

AFTERTHOUGHTS

One cannot discharge one's duty by making a monumental paper structure and then not implementing it. — A counsel during the trial following the Longford explosion (see Section 4.2)

At every safety conference, speakers describe their safety management systems. I often wonder how well they are implemented. Descriptions of their company's accidents might tell us more.

I remember the first time I rode a public bus . . . I vividly recall the sensation of seeing familiar sights from a new perspective. My seat on the bus was several feet higher than my usual position in the back seat of the family car. I could see over fences, into yards that had been hidden before, over the side of the bridge to the river below. My world had expanded. — Ann Baldwin, *Biblical Archaeology Review*, May/June 1995

We need to look over fences and see the many opportunities we have to learn from accidents.

Before Columbus made his discovery the Spanish Royal family believed the Straits of Gibraltar to be the last outpost of the world. Their coat of arms depicted the Pillars of Hercules, the Straits of Gibraltar, with the motto *Nec Plus Ultra* (No More Beyond). After Columbus set sail the Royal family, with great economy, did not change their coat of arms. They merely erased the negative so that their motto now read *Plus Ultra* (More Beyond). — Danny Abse, *Goodbye, Twentieth Century*

At this point I bring my work to an end [and leave others to go beyond]. If it is found well written and aptly composed, that is what I myself hoped for; if cheap and mediocre, I could only do my best. For just as it is disagreeable to drink wine alone or water alone, so the mixing of the two gives a pleasant and delightful taste, so too variety of style in a literary work charms the eyes of the reader. Let this then be my final word. — The ending of *2 Maccabees* (early 1st century BC).

Index

Printed and bound by CPI Group (UK) Ltd, Croydon, CR0 4YY

03/10/2024

01040430-0004